Knowledge and Colonialism: Eighteenth-century Travellers in South Africa

Atlantic World

Europe, Africa and the Americas, 1500–1830

Edited by

Wim Klooster
Clark University

and

Benjamin Schmidt
University of Washington

VOLUME 18

Knowledge and Colonialism: Eighteenth-century Travellers in South Africa

By
Siegfried Huigen

BRILL

LEIDEN • BOSTON
2009

On the cover: "Coba Caffer Captein" (Gordon Atlas, G75). Courtesy of the Rijksprentenkabinet, Amsterdam.

This book was originally published as *Verkenningen van Zuid-Afrika. Achttiende-eeuwse reizigers aan de kaap* (2007).

This book is printed on acid-free paper.

Library of Congress Cataloging-in-Publication Data

Huigen, Siegfried.
Knowledge and colonialism : eighteenth-century travellers in South Africa / by Siegfried Huigen.
p. cm. — (Atlantic world : Europe, Africa, and the Americas, 1500–1830 ; v. 18)
Includes bibliographical references and index.
ISBN 978-90-04-17743-7 (hbk. : alk. paper) 1. Cape of Good Hope (South Africa)—Description and travel. 2. Cape of Good Hope (South Africa)—Description and travel—Sources. 3. Travelers—South Africa—Cape of Good Hope—History—18th century. 4. Europeans—South Africa—Cape of Good Hope—History—18th century. 5. Ethnology—South Africa—Cape of Good Hope—History—18th century. 6. Ethnological expeditions—South Africa—Cape of Good Hope—History—18th century. 7. South Africa—History—To 1836. 8. South Africa—Colonial influence. 9. South Africa—Description and travel. 10. South Africa—Description and travel—Sources. I. Title. II. Series.

DT2020.H85 2009
968.03—dc22

2009017888

ISSN 1570-0542
ISBN 978 90 04 17743 7

PRINTED IN THE NETHERLANDS

CONTENTS

PREFACE

South Africa played an important role in the formation of an image of Africa in Europe during the 'long eighteenth century'. The deep interior of the country was accessible only from the Cape. Except for a few settlements along the coast, the rest of Africa south of the Sahara was a closed territory to Europeans until the nineteenth century. This book analyses how Southern Africa was represented in words and images during the period when the Cape Colony was under the jurisdiction of the VOC (Dutch East India Company) and the Dutch state (1652–1814).

Southern Africa was explored by travellers of various nationalities and with various objectives. Besides the *landreizigers* (overland travellers) employed by the VOC, who explored only the economic potential of the interior, scientific travellers also arrived in South Africa in the eighteenth century. These men—most of them from other countries than the Netherlands—came looking for answers to questions that occupied European scholars. Everything between heaven and earth attracted their interest. They usually published their findings, and in this way they sometimes reached a large audience.

In this book I have generally emphasised the ethnographic representations in travel accounts and in images. All travel accounts have something to say about the culture and the appearance of the indigenous population. The extent of the ethnographic description could range from incidental remarks to, in exceptional cases, a complete book on an indigenous population group such as *De Kaffers aan de Zuidkust van Afrika* (The Kaffirs on the South Coast of Africa), written by the Dutch army officer Ludwig Alberti in 1810. I also refer to zoological and cartographic representations in passing.

The full manuscript of the original Dutch version of this book, which was published in 2007, was read by Ena Jansen, Jean Kommers and Helize van Vuuren. Adriënne Zuiderweg and Kees Rookmaaker read some chapters. Herman Roodenburg made valuable comments on an earlier version of chapter 5 that had appeared as an article in the *Nederlands Kunsthistorisch Jaarboek*; Marijke Meijer Drees did likewise with chapter 8, which had been published in a different form in *Nederlandse Letterkunde*. The comments of these readers were very helpful. Jean Kommers in particular very generously supplied additional information and comments. Many other people also offered assistance during the research for and preparation of the manuscript. To all of them my sincere thanks. I also thank Tony Moen for the translation into English and Stellenbosch University for providing funds for making the translation possible.

Stellenbosch, February 2009

LIST OF ILLUSTRATIONS

CHAPTER ONE

WAGON ROUTES: AN INTRODUCTION

The Heerenlogement

Some 250 km north of Cape Town there is a cave where more than a hundred names have been carved into the rock walls. The oldest inscriptions date back to the start of the eighteenth century. The cave was used as a stopover for travellers who were on their way from Cape Town to the northwestern parts of the present South Africa and to Namibia. It soon received the name "Heerenlogement", or Gentlemen's Lodge, a word used in the seventeenth and eighteenth century in the Netherlands for the better class of accommodation.[1] Giving such a name to a bare cave on the edge of a sandy plain was probably meant as a joke, which is why travellers sometimes referred to the "so-called" Heerenlogement (figure 1). Under his drawing of the Heerenlogement, the Dutch traveller Robert Jacob Gordon (1743–1795) wrote in 1779: "so-called gentlemen's lodge 3 miles south of the Olifants River on the farm of Pieter van Zyl" with a note at "gentlemen's lodge": "being a cave in a cliff".[2]

This cave was declared a national monument early in the twentieth century. To reach it, you must leave the gravel road that passes it and climb up a winding footpath through low shrubs. The "cave in a rock" is actually more an overhanging rock resembling an artillery bunker of Hitler's Atlantic Wall. The walls are indeed covered with travellers' names: "Slotsboo 1712", "Rhenius 1726", "F. Vailant 1783" (fig. 2). The tree about which eighteenth-century travellers already wondered how it stayed alive in this arid environment still grows from a crack in the rock. Standing at the mouth of the Heerenlogement cave, you have a magnificent view of the Sandveld, a plain between the mountain ridge

[1] For the meaning of 'Heerenlogement', see *Woordenboek der Nederlandsche Taal op CD ROM* (2000).

[2] Gordon's drawing is included in *Robert Jacob Gordon, Cape Travels, 1777 to 1785* (Raper & Boucher, 1988). For the list of names on the walls of the Heerenlogement see Kerby (1942).

in which the Heerenlogement is located and the nearby cold Atlantic Ocean in the west. As I was climbing up the bunker-like roof of the cave, a bank of fog was rolling in from the sea. Within a few minutes the temperature seemed to have dropped several tens of degrees. Suddenly I could see my own breath.

From the early days, the Heerenlogement was a landmark on the route from Cape Town to the north. Because of this status, a traveller simply had to drop in—there was also a fountain at the foot of the mountain in which the cave is—and perhaps carve his name in the rock or make a drawing. Robert Jacob Gordon and the Surinamese-French traveller François le Vaillant (1753–1824) made drawings of the Heerenlogement. In his *Second Voyage* (1795) Le Vaillant also describes his disillusionment after a visit to the cave. The site did not meet his high expectations of finding an Arcadian grotto decorated with murals and lush trees at a crystal spring.

The landscape consisted of bare mountains, the spring was muddied after his arrival by his oxen and by his "Hottentots" and the trees were limited to the tree mentioned earlier growing from a crack above the cave. Moreover, when Le Vaillant wanted to go to sleep in the "natural hall" of the cave, his peace was disturbed by hundreds of wood pigeons which had settled down in the tree above him. The rich frescos turned out to be nothing more than two "caricatures" of an elephant and an ostrich and the names of travellers. His grumpy description of the cave is that of someone who has not closed an eye all night.[3] Nevertheless, Le Vaillant added his name—"F. Vailant 1783"—to the names of his predecessors.

In another sense, however, the name Heerenlogement is quite applicable. The travellers who set out on journeys of discovery in the eighteenth and nineteenth centuries and carved their names on the walls of the cave as they made their way north were indeed gentlemen. During the period when the Cape Colony was officially governed by the Dutch (1652–1814, including periods of English occupation) there was only one woman who left behind an account of her travels: Augusta de Mist (1783–1832), the daughter of the Batavian commissioner-general Jacob Abraham de Mist (1747–1823). In 1803 she accompanied her father on

---

[3] Le Vaillant (1795, pp. 111–113).

a long journey through the interior of which a short description was published in the woman's magazine *Penelopé* years later.[4]

Descriptions of Southern Africa

The subject of this book is the attempts made by travellers in the period during which South Africa was governed by the Dutch to describe the interior and the indigenous inhabitants of Southern Africa. These travellers—mostly from the eighteenth century—were a motley crowd. The expeditions of the Dutch East India Company (Verenigde Oost-Indische Compagnie, VOC) were primarily concerned with the economic potential of the interior; at the beginning of the nineteenth century Dutch government officials wanted to gather facts about the unfamiliar regions that they had to govern. It was also in the eighteenth century that scientifically trained individuals for the first time travelled into the interior of South Africa with the purpose to open up the foreign world for the benefit of European science. What all travellers had in common was that they wanted to establish the exact facts; however, in accumulating knowledge the scientific travellers lacked an instrumentalist objective. This distinguished them from the VOC travellers, who had to explore the economic potential of the interior to the exclusion of almost everything else. In the following chapters, the main focus will be on the scientific exploration of Southern Africa and its inhabitants, because this category produced the greatest achievements. The accounts produced by the VOC are materially different. The objectives of VOC expeditions were limited, with the result that their representations are rather meagre compared with the work of scientifically trained travellers. However, for a multi-faceted picture of the exploration of South Africa they remain indispensable.

The oldest texts about South Africa are travel accounts produced by world travellers. The Cape was described by the "Callers at the Cape", as Raven-Hart called them, from the first moment they set foot on shore around 1500.[5] These were European travellers from many countries, including the Netherlands, who stopped over briefly at the Cape to collect water and obtain meat from the Khoikhoi or Hottentots.

---

[4] De Mist (1835, pp. 71–127); see also chapter 8.
[5] Raven-Hart (1971).

These short visits gave rise to stereotypical descriptions of the Cape, the Khoikhoi and wild animals, with Table Mountain in the background. Seen through European eyes, it was an unattractive region; the landscape was inhospitable and the inhabitants were uncivilised. Unlike Asians, the Khoikhoi did not come up to European standards. Neat clothes or a technologically developed, material culture were nowhere to be seen. Their language sounded like the gobbling of turkeys.

The establishment of a colony partly changed this discourse. Partly, because even after 1652 the Cape remained a regular element of the accounts of journeys to and from Asia, and most authors followed the example of their predecessors in their descriptions. Visitors therefore simply reiterated the stereotypical descriptions, which were soon expanded by equally standardised accounts of the new developments that were taking place in the colony: the VOC's gardens at the foot of Table Mountain, the 'Vlek' (i.e. 'hamlet', the settlement that would be called Cape Town after 1800) and the towns in the interior, especially Stellenbosch. Really new descriptions were initially only to be found in the writings of people in the service of the VOC who undertook expeditions into the interior at the behest of their principals. Only in exceptional cases were the descriptions of these overland journeys published in the seventeenth and eighteenth centuries, as in the multivolume *Oud en Nieuw Oost-Indiën* (1724–1726) by François Valentyn (1656–1727).

In the second half of the seventeenth century several exploratory journeys were undertaken into the interior. Some of these expeditions had set themselves the ambitious goal of mapping out the economic potential of the interior. Initially, one aid to this end was the very dubious information provided in the *Itinerario* of 1596 by Jan Huygen van Linschoten (1563–1611). In this book Van Linschoten published all the information he had collected about the secret trade networks of the Portuguese in Asia. For the VOC the *Itinerario* was at first a useful guide enabling it to determine with whom advantageous trade relations could be established in Asia and which strategic points would allow them to control trade in the Indian Ocean area. Van Linschoten also dedicated a number of pages to Southern Africa, in which he stated among other things that the Portuguese had found gold in Monomotapa. Because of Van Linschoten's reports about gold, plans were drawn up not long after the establishment of the colony at the Cape to send expeditions to Monomotapa. Van Linschoten had even added a map to his book indicating the location of the "empire" of Monomotapa. From 1660 the VOC sent expeditions inland based on this information.

The successive VOC travel accounts of the seventeenth century read like a serial in which a consistent representation of the interior gradually develops. Conspicuous in this representation of the South African interior is the combination of credulity and empiricism. On the one hand the new information led to corrections of the largely fictitious maps produced by Van Linschoten. On their way, the travellers described their routes, which were then processed into empirically justified, more reliable maps. On the other hand the fictitious representations of the interior were not quite rejected: they simply shifted farther inland. On the known side of the frontier lay the area inhabited by the uncivilised "Hottentots", which was described in negative terms. Beyond the frontier, the existence of highly developed cultures could not be ruled out. At the end of the seventeenth century, however, the region inhabited by uncivilised peoples already extended up to the Namaquas in the north-west, in the region of the Great River (Orange River), because the similarities between the Namaquas and the inhabitants of the coastal strip around Table Mountain were unmistakable and because the Namaquas were no longer as well disposed towards the Dutch as they used to be. Hostile stereotyping played a part in the conceptualisation: from the moment the Namaquas adopted a more antagonistic stance, they were described as "savages".[6] Despite the disappointing results of the searches for Monomotapa, the VOC continued to expect to make contact with this "empire" until the eighteenth century (see chapter 3).

Most of the VOC expeditions lacked scientific objectives.[7] These were introduced into the European discourse on South Africa by the German scientist Peter Kolb (1675–1726). The publication of Kolb's *Capvt Bonae Spei Hodiernvm* ('The Current State of the Cape of Good Hope') in 1719 was foundational to the development of a scientific and scholarly discourse on South Africa. Kolb, who lived in Cape Town and Stellenbosch between 1706 and 1713, wrote a systematically structured encyclopaedic work of 900 folios in which he described the geography, nature, indigenous population and the colonial settlement at the Cape. Until deep into the eighteenth century, this was the most authoritative book on Southern Africa. Kolb's exclusive and comprehensive coverage

[6] Huigen (1996a, pp. 23–60).

[7] Only in exceptional cases, such as the expedition to Namaqualand led by Simon van der Stel, were real efforts made to collect scientific facts. Even in this case, economic considerations were the main object. See Huigen (1996a, p. 33).

of the Cape was a novelty at the beginning of the eighteenth century; for earlier visitors the Cape had never been more than an excursion of a few pages, after which they quickly moved on to Asia or back to Europe. Moreover, through his caustic criticism he rendered the seventeenth-century publications about the Cape irrelevant for travellers coming after him, so that after Kolb little reference is made to the work of earlier authors.[8] Until the end of the eighteenth century, Kolb's book remained the key point of reference. Furthermore, Kolb introduced the critical examination of existing representations of South Africa, which he compared with his own experience based on 'autopsy', i.e. observation with his own eyes. After Kolb this became a permanent element of the representations of South Africa. Of course this is not due to him only; it is also part of the epistemology of the eighteenth-century empiricism, which strongly emphasised personal observation.

After the publication of Kolb's book more monographs on South Africa appeared during the last quarter of the eighteenth century, culminating in the travel accounts of the Surinamese-French ornithologist François le Vaillant and the Englishman John Barrow (1764–1828), which stand out by their scope, quality and wide distribution. The scientific status of travel accounts of this kind was generally high. For example, the plant collector of the Royal Botanical Gardens in Kew, Francis Masson (1741–1805), published an account of his South African journeys in the *Philosophical Transactions of the Royal Society of London*, arguably the most important European scientific journal of the seventeenth and eighteenth century.[9] Although Masson had no thorough scientific training, his observations were nevertheless considered to be of scientific interest. A similar view was taken of the travel journal of a VOC expedition of 1761–1762 (cf. chapter 4).

This scientific interest in Southern Africa was nourished by a new empirical science that was interested mainly in accumulating and evaluating an encyclopaedic collection of empirical facts.[10] The natural science of the eighteenth century was characterised by a "factual sensibility", a predilection for facts.[11] Unlike in the older Aristotelian empiricism, where facts had to help make theories plausible, 'bare' facts were

---

[8] The older publications remained important, however, for authors who did not occupy themselves solely with South Africa.

[9] Masson (1776).

[10] Bourguet (1996, pp. 289–293).

[11] Daston (1988, pp. 452–470).

studied for their own sake in the new natural science and facts formed the basis of theories instead of illustrating them. What mattered were unrelated pieces of experiential knowledge, all of which were equally valuable in their own right.[12] For exotic facts, eighteenth-century science relied heavily on travel accounts, which therefore enjoyed the same epistemological status as the results of laboratory experiments.[13] This explains why Francis Masson, a simple collector of plants, was able to publish his travel account in the *Philosophical Transactions* in 1776.

The nature of the facts which were to be investigated differed in many respects from what can be regarded as the field of scientific study today. In the first place, there was no specialisation.[14] Everything between heaven and earth could expect to attract the attention of the scientific traveller,[15] even if the training of some of them, especially the pupils of Linnaeus such as Anders Sparrman (1748–1820) and Carl Peter Thunberg (1743–1828),[16] displayed a distinct preference for zoology and botany. It is not really surprising that the geographical length of the Cape of Good Hope was investigated, the exact height of Table Mountain was determined, weather patterns were measured, anatomical studies of animals were done or words from indigenous languages were recorded. But scientific interest was also aroused by 'monstrous' phenomena which are stored in formaldehyde in the basements of museums of natural history to this day.[17] For example, there was great (not only voyeuristic, but also scientific) interest in the elongated labia of Khoi women (fig. 3)—did all Khoi women have such an "apron", and if so, was this anatomical peculiarity induced artificially or was it a congenital defect? During his visit to the Cape, James Cook (1728–1729)

---

[12] Daston (1988) writes especially about the sixteenth and seventeenth century, but her statement is also applicable to the eighteenth. See Dear (2001, pp. 1–9).

[13] Carey (1997).

[14] The eighteenth century saw little more than the first steps towards the development of, for example, geography, linguistics and anthropology as scientific disciplines (cf. Vermeulen, 1996).

[15] Stafford (1984, pp. 43, 59).

[16] I always refer to the modern English editions of the travel descriptions of Sparrman (1975–1977) and Thunberg (1986) in the Van Riebeeck series. The original text of Sparrman's *Resa till Goda Hopps-udden, Södra pol-kretsen, och omkring jordklotet samt till Hottentott- och Caffer-landen åren 1772–76* ('A Voyage to the Cape of Good Hope towards the Arctic Circle round the World and to the Country of the Hottentots and the Caffres from the Year 1772–1776') appeared in Stockholm in 1783; Thunberg's text appeared as part 4 of *Resan uti kejsardömet Japan, på Java och Ceilon samt hemresan* in Uppsala in 1793.

[17] Daston (1991), Daston & Park (2001).

refers to the "Hottentot apron" as "the great question among natural historians" and in The Hague Diderot (1713–1784) questioned Robert Gordon about his experiences in this regard.[18] In his travel journals, the ever sober Gordon mentions gynaecological examinations four times. Indeed, the women examined were found to have such an "apron" (cf. chapter 5).[19] But it went even further. As there was always something new to be discovered in the unknown Africa, phenomena of which the existence had long been doubted in the scientific discourse might well be found in Southern Africa. Scientific curiosity had such an appetite for sensational discoveries in South Africa that it raised the occasional learned brow in Europe. I will give an example of this in chapter 10.

The Cape Colony and its hinterland

In the eighteenth century, South Africa offered something of scientific interest to everyone. The Swede Thunberg, for example, embarked on the scientific description of the luxurious South African flora; his compatriot Sparrman and the Dutchman Gordon collected information on the fascinating fauna at a time when all of South Africa was one large safari destination; Le Vaillant collected birds and produced impressive ornithological atlases after his return. Everybody was doing ethnography and trying to put his route on the map. In between there were occasional critical comments on the repressive colonial regime. One consequence of the relative prominence of the Cape in European descriptions of Africa was that the image that had been formed from a handful of descriptions of South Africa became metonymic for all of Africa. A broader perspective was obtained only in the nineteenth century, after explorers had crossed the dark heart of Africa.[20]

---

[18] Cook, quoted from Schrire (1996, p. 347). Diderot describes his meeting with Gordon in his *Voyage en Hollande* (1819); for the discussion of this meeting see Porter (1819, p. 82); Cullinan (1992, pp. 22–24).

[19] The interest in the "apron" continued into the nineteenth century (see Qureshi, 2004, pp. 243–245, 247).

[20] Before 1800 only the European knowledge of the Nile Valley could be compared to that of South Africa (Heawood, 1969, p. 390). Although Portuguese explorers had already travelled deep into Southern Africa from the Mozambican side during the sixteenth century, these accomplishments were largely forgotten during the seventeenth and eighteenth century. Accordingly, these Portuguese exploits left no traces in the eighteenth century European discourse on Africa. Noteworthy are also Adanson's travels in Senegal in 1749–1753 (published in Adanson 1757).

The reason for the great scientific interest in Southern Africa during the eighteenth century was a consequence of the limitations that curtailed European colonial expansion elsewhere in Africa. Until the end of the eighteenth century the interior of Africa south of the Sahara remained largely inaccessible due to disease, impenetrable forests and hostile political structures. The establishment of a settlement at the Cape of Good Hope and an expansion of the sphere of colonial influence in the eighteenth century made South Africa the only part of sub-Saharan Africa where one could travel deep into the interior. The open landscape—large parts consist of semi-desert—and the absence of tropical diseases also helped. For the latter reason, the English colonial official John Barrow had suggested that the Cape was indispensable to the British Empire. Here large numbers of troops could be kept on standby in good health for deployment in the infernal climate of Asia.[21]

Owing to the early colonisation after 1652 South Africa also offered the traveller a rudimentary infrastructure that was absent in the rest of Africa south of the Sahara. At the Cape, an area the size of Spain had been colonised by the end of the eighteenth century (fig. 4). In Cape Town the traveller could equip himself for an expedition. He could get ox-wagons, gunpowder, trading goods, food and drink. He could also engage experienced personnel—guides, drivers, "Hottentots"—and obtain advice from local experts such as the military commander Robert Jacob Gordon, who had visited all the corners of the colony. Le Vaillant gives us some idea of the equipment for such an expedition. On his first journey in 1781–1782 he had two ox-wagons with 30 oxen, three teams of 10, of which one was always kept in reserve. The first wagon was the "master wagon", Le Vaillant's sleeping quarters and laboratory on wheels provided with boxes in which the specimens he collected on his way could be stored. This wagon also carried the boxes of gunpowder. Le Valliant often slept on a mattress in the wagon on top of the boxes with specimens and gunpowder. The second wagon was the kitchen and tool wagon, with kitchen equipment, stocks of coffee, sugar, brandy and tobacco. Tools were also transported on the wagon, including tools for making bullets for his sixteen flintlocks and double-barrelled pistols. He also had a large tent—shown on many illustrations—and a smaller "soldier's tent" (fig. 5). A varying number of people accompanied him

[21] Barrow (1804, pp. 158–168).

on his journey; at one stage there were as many as forty.[22] In order to feed all these people he regularly had to go hunting.

From 1652 until 1814 the Cape Colony was officially governed by the Dutch. In 1652 the VOC had established a refreshment post for its ships at the Cape of Good Hope. The limited purpose of this refreshment post was to reduce the large number of deaths on board of the VOC's sailing ships. For this reason a small settlement was established at the tip of Africa with a fort that could defend the anchorage against European rivals and with gardens at the foot of Table Mountain. In 1652 several hundred Europeans were living in this colony, subsequently supplemented with slaves. In the course of the eighteenth century the population of the Cape Colony—colonists, slaves and Khoikhoi—expanded to about 60,000 in 1798. All these people could assist travellers.

If it had been up to the VOC, the colony would have remained confined to a small part of the Western Cape—from the coast of Table Mountain and False Bay up to the mountains that close off the interior. The intention was to appropriate no more land than was necessary to provide the ships calling in Table Bay with food. However, after 1657 the VOC needed the help of independent colonists to achieve this goal, because the nomadic Khoikhoi did not practise agriculture and could not supply enough cattle for slaughter either. Such colonists were called freeburghers. These colonists caused the colony to expand further, especially in an eastern direction. In the course of the eighteenth century, the colony had become so large that even within its borders there was enough for travellers to discover. On the northern and eastern periphery, the more densely populated parts in the vicinity of Cape Town spilled over into an open frontier area where colonists and indigenous inhabitants lived side by side until the end of the eighteenth century.[23] Although the Fish River was declared the official eastern border in 1778 during the term of governor Joachim van Plettenberg (1774–1785), indigenous groups hardly paid any attention to this. In the north the demarcation remained unclear even longer, because there was no strong opponent there like the Xhosas in the east.[24] The region inhabited by Europeans in the north-east even shrank a little towards the end of the

[22] Le Vaillant (1790, pp. 49–52).

[23] In specialist historical studies, this is referred to as an "open frontier zone" (Giliomee 1981, pp. 76–119; 1992, pp. 421–471); Legassick (1992, pp. 358–420).

[24] Penn (2001, pp. 19–46).

eighteenth century due to the persistent guerrilla war waged against them by the San (Bushmen). The colonial frontier was populated by, among others, men who had moved out beyond the colonial authority and often lived together with indigenous women.[25] These freebooters could be useful to travellers because they were able to obtain provisions or act as guides. The border was closed only in the nineteenth century, thus depriving the indigenous groups in the colony of their means of subsistence, so that they lost their independence.

In the period between 1652 and 1814 the government of the Cape Colony changed hands a few times. The VOC remained at the helm until 1795. After the Dutch Republic had been occupied by French troops in 1795 and the stadtholder had fled to England, an English force landed at Muizenberg south of Cape Town. Due to political reluctance of the Dutch administrators who had remained loyal to the stadtholder—they did not really want to defend the Cape on behalf of the newly established, revolutionary Batavian Republic in the Netherlands—the English captured Cape Town without too much trouble. The interim English administration remained until 1803. This was because in 1802 it was agreed at Amiens that the Cape Colony would be returned to the Batavian Republic. This Batavian administration lasted from 1803 to 1806, when a large English force attacked a mixed Batavian army consisting of Dutch soldiers, French sailors, Cape Malay gunners, Khoikhoi infantry, German mercenaries and Cape citizen cavalry at Blaauwberg, north of Cape Town. Only when the Convention of London was signed in 1814 was the Cape Colony, together with other former Dutch possessions, officially transferred to the English.[26]

Under the English the colonial regime in South Africa changed. The VOC and the Batavian Republic lacked the capacity to exercise authority in the border areas. However, the English regime could allow itself the liberty of a confrontation with unwilling colonists and the Xhosas on the eastern border. Consequently the porous border hardened into

---

[25] See in this regard Penn (2005, pp. 164–165, 170–217).

[26] The text of the Convention of London can be viewed on the Internet: "Convention between Great Britain and The Netherlands relative to the Dutch colonies; trade with the East and West Indies; Etc. signed at London, August 13, 1814" (http://www.guyana.org/Western/1803%20to%201840.htm). In South Africa it is often said that the Cape was sold to England by the Netherlands. However, there is no mention of this in the convention. Reference is only made to English financial support for strengthening the Dutch defence against France, which of course was in the interest of England.

a demarcation line between the colony and indigenous regions. In the nineteenth century the Xhosas were driven beyond the official borders of the colony and ultimately subjected to English authority. In other respects, too, the 'English nineteenth century' differed from the preceding Dutch period. On the one hand, missionaries succeeded in persuading the English regime to follow a humanitarian policy that aimed to improve the position of the indigenous population and slaves in the colony. On the other hand, ethnic differences became more acute and the colonial wars became bloodier.[27]

While the Cape was under Dutch administration, Dutch was also the language of government and the most generally used language of communication. The Swedish traveller Thunberg, whose final destination was Japan, had even come to the Cape to learn Dutch here. In Japan Dutch was the only language for communicating with the outside world.[28]

The colonial expansion in South Africa did not benefit scientific travellers in all respects. On the one hand, it did make travelling a good deal easier because European colonists had settled here and there in the interior and primitive connecting routes had been established. For example, you knew that you could find water and shelter at the Heerenlogement. Survival in the wild could depend on such knowledge. However, the drawback was that especially the traveller with ethnographic interests—who will receive a fair amount of attention in this book—found it increasingly difficult to find indigenous inhabitants who had not been influenced by the dominant colonial Dutch culture.[29] Especially Khoikhoi whose culture had remained unaffected were more and more difficult to find at the end of the eighteenth century. Those who lived in the south-west had been decimated by a smallpox epidemic in 1713 or had been absorbed in the colonial economy as labourers.[30]

---

[27] Keegan (1996).

[28] During Japan's period of isolation from the outside world (1641–1853) the VOC's trading post on the small island of Deshima (also spelled Dejima) near Nagasaki was the only connection between Japan and the outside world.

[29] Compare the following remark by Gordon: "om over de Sneeuwberge in het Land der zogenaamde Wilde *Bosjesmans-Hottentotten* te trekken, dewijl men thans in onse Colonie schoon hoezeer uitgestrekt, weynigh aanmerkenswaardigs te zien krijgt" (to trek across the 'Snow Mountains' into the Country of the so-called Wild *Bushmen-Hottentots*, while at present in our Colony, large as it may be, little that is worthy of note is to be seen) (copy of letter by Gordon to Fagel, Brenthurst Library, Johannesburg, Gordon Papers, 1779, MS107/13, "Lettres écrites a differentes Personnes").

[30] Szalai (1983).

Many of their original customs had therefore been lost. Large numbers of San (Bushmen) had been killed during punitive expeditions. For ethnographic research, it was therefore necessary to go deeper and deeper into the interior, up to the outskirts of the colony or even farther.

Travellers and their accounts

The fact that the Cape was under Dutch administration between 1652 and 1814 did not mean that it was mainly the Dutch who described the interior. It is true that the VOC sent expeditions inland, but the reports of these expeditions were usually not published. This meant that, for example, no use was made of the unique information that VOC expeditions had collected in the north-east of South Africa and in Mozambique (chapter 3). The main scientific contributions to European knowledge of South Africa were made by a cosmopolitan group of travellers, some of whom will be dealt with in the following chapters.[31] The only significant Dutch traveller in this group was the VOC army officer Robert Jacob Gordon. The international composition of the group of travellers that crossed the South African interior was not unusual; the travellers who explored South America and Siberia during the same period were usually not Spaniards or Russians either.[32]

The exploration of South Africa by this cosmopolitan company of scientifically trained travellers was not coordinated by what the anthropologist of science Bruno Latour has called "centers of calculation", in other words centres where the accumulation of geographical, natural historical and anthropological knowledge was concentrated and that

[31] Although Sparrman and Thunberg—besides Kolb, Gordon, Le Vaillant and Barrow—are also important, I will not dedicate a chapter to them because I cannot read Swedish and it is irresponsible to do a close reading of a text in translation. In this introduction and in the final chapter, in which I want to reveal general connections between the travel accounts, I do think that I can make use of translations of their work. In the following I always refer to the English translations of their work. As regards the travel accounts which are relevant here, I follow the broad definition of these early modern accounts given by Carey (1997, pp. 279–280): "[...] we should not assume that a hermetic genre of travel writing exists, discretely defined. Narrowly conceived, the genre consists of first-person relations of experience accrued in the course of travel. More broadly, one must include in the category second-hand summaries of the journeys undertaken by others, and 'histories' of regions."

[32] Bourguet (1996, pp. 294–296). Early modern travel descriptions were also cosmopolitan in the sense that they were conditioned by pan-European factors (cf. Rubiés, 2000, p. xii).

had the authority to discipline individuals and institutions across a broad spectrum.[33] Individual travellers did maintain contact with such centres, such as Linnaeus in Uppsala, Joseph Banks in London and Professor Allamand in Leiden, but in the travel accounts these were of subordinate importance, for instance in the use of the Linnaean nomenclature.

Despite the absence of a distinct scientific centre of calculation, there was certainly an accumulation of knowledge that was accessible to a broad group of readers in the form of published travel accounts. The wide range of this knowledge appears among others from a remark by Hinrich Lichtenstein (1780–1857), the author of *Reisen im südlichen Afrika in den Jahren 1803, 1804, 1805 und 1806*,[34] who had lived in South Africa for three years, that "all of Europe that can read" was familiar with the account of John Barrow's travels that had appeared in 1801.[35] Translations enabled the European public to compare the accounts written by different authors.[36] In one of his reviews of South African travel texts, the German world traveller and author of travel accounts Georg Foster (1754–1794) compared *A narrative of four journeys into the country of the Hottentots* by William Paterson (1755–1810) with reports that had appeared earlier and assumed that the reader of his review was aware of this as well ("man kennt schon"—one already knows).[37] The accumulation of knowledge appears most clearly from the travel baggage of the Batavian commissioner-general De Mist (no scholar, but a government official) and Hinrich Lichtenstein on their journey through the interior in 1805. Their suitcases contained the most important publications about South Africa to date: Lichtenstein had brought along the accounts by Peter Kolb, Anders Sparrman, Carl Peter Thunberg, François le Vaillant and John Barrow; in the suitcases of De Mist there were Stavorinus, Crossigny and the French translation of Barrow by Degrandpré.[38] This allowed them to compare their observations on the

[33] Latour (1987, pp. 215–257), Miller (1996, pp. 21–37). These centres can assume a variety of forms: museums, learned societies, trading companies and also persons such as Joseph Banks in London, Buffon in Paris and Linnaeus in Uppsala.

[34] 'Travels in South Africa in the years 1803, 1804, 1805 and 1806'.

[35] Lichtenstein (1811, p. 617).

[36] Translations made texts accessible to different language areas (Osterhammel, 1998, pp. 29–30)

[37] Forster (1992, p. 190).

[38] Lichtenstein (1811, p. 27). The books in question were: *A Voyage to the Cape of Good Hope towards the Arctic Circle round the World and to the Country of the Hottentots and the Caffres from the Year 1772–1776* (Sparrman, 1974–1977); *Travels at the Cape of Good*

ground with what their predecessors had written and to add their own description to what was already available. Even while travelling, the traveller thus remained a member of the European "communication community" with which debates were held on the worlds and people observed.[39] In the case of South Africa, this accumulation of knowledge was nothing new. An accumulation of geographical knowledge on Southern Africa already existed since the second half of the seventeenth century in the network of the VOC, but this reached the public at large only incidentally (chapters 3 and 4).[40]

What was new was the scholarly traveller journeying far into the interior. Early in the eighteenth century Peter Kolb lived in Cape Town and Stellenbosch for a number of years, from where he had made excursions inland, and later the abbé Louis de la Caille (1713–1762) travelled up to St Helena Bay north of Cape Town. The generation of scientific travellers at the end of the eighteenth century, starting with Anders Sparrman, penetrated more deeply into the interior on their own initiative. Journeying into the interior was their main reason for coming to 'Africa'. These travellers frequently crossed the (indistinct) borders of the colony to personally observe unfamiliar natural phenomena and peoples who had not been subjected to Western influence, or at least they attempted to create the impression that the descriptions of remote regions and peoples were based on observation.[41] The demand for scientific knowledge was great and the travel accounts preferably had to be based on personal observation.[42]

The scientific travellers could profit from various kinds of patronage.[43] Gordon, Barrow and Lichtenstein had the opportunity to visit remote parts within and beyond the colony in their official capacity. The Englishmen Masson and Paterson received financial support to supplement the collections of the botanical gardens of Lady Strathmore and of the English king in Kew. The other three, Thunberg, Sparrman and

Hope (Thunberg, 1986); *Voyage de M. le Vaillant dans l'intérieur de l'Afrique par le Cap de Bonne-Espérance. Dans les Années 1780, 81, 82, 83, 84 & 85* (Le Vaillant, 1790); *Second Voyage dans l'intérieur de l'Afrique, par le Cap de Bonne-Espérance, dans les années 1783, 84 et 85* (Le Vaillant, 1795).

39 Harbsmeier (1994, p. 23).

40 Huigen (1996a, pp. 23–60); Huigen (1996b).

41 According to Forbes (1973, pp. 51–52, 90–91) Le Vaillant did not visit all the places he described.

42 Blanke, 1997, I, 187.

43 Sparrman was the only one who had to make do on its own; he had to make a living as a private tutor and doctor in the Cape Colony.

Le Vaillant, depended on the generosity of several sponsors. Le Vaillant was able to travel to the Cape with the patronage of Jacob Temminck, a director of the VOC. However, when he lost all his possessions during a naval battle in Saldanha Bay in the Fourth English War he was generously assisted by the VOC official Boers, to whom he later dedicated his *Voyage*. For travel advice Le Vaillant was able to consult Colonel Gordon, commander of the VOC Garrison at the Cape. The more sedentary Kolb was initially subsidised by a German baron during his stay in Cape Town. When the financial support ended after the baron's death, he entered the service of the VOC as secretary of the inland district of Stellenbosch and Drakenstein. The latter capacity in particular gave him access to information that he subsequently used for his major work on South Africa, *Capvt Bonae Spei Hodiernvm*. What all travellers had in common was that they received assistance from the colonial authorities in one way or another. Even those who were not officials were allowed to travel freely in the interior, even though the VOC imposed restrictions on the Englishman Masson when his English colleague Paterson came under suspicion of having spied for England.

To be taken seriously in Europe, the traveller had to have a thorough scientific education. Texts produced by employees of the VOC and colonists, such as Jacob van Reenen's report of a journey to "Caffraria" (the Eastern Cape) in 1790–1791, no longer met the requirements at the end of the eighteenth century. Georg Forster said that the report contained nothing new and was merely a dry summary of the places and rivers visited, the daily distance covered and the problems encountered on the way.[44] This is indeed an accurate description of this report, and Forster's characterisation is also applicable to the travel journals of most of the other travellers who had little training. Generally speaking, there is a considerable difference between the writings of VOC travellers on the one hand and those of the scientifically trained travellers on the other. The work produced by the first group is not much more than a diary in which the travel experiences are recorded. The information about the foreign world is quite thin in these journals, because there is no broader perspective. In order to give texts of this kind some scientific importance, they were sometimes supplemented by a scholarly editor (chapter 4). The scientific travellers, however, saw the surrounding real-

[44] Forster (1992, p. 363). This was a report edited by the English captain Riou (Van Reenen, 1792).

ity from a broader perspective, because in principle everything could be relevant to contemporary science.

The publications of the scientifically trained travellers were also based on travel journals, but they took much trouble editing their notes when they published their experiences. In Le Vaillant's case even editors were engaged.[45] As regards the format of the published travel account, one had a choice between two main types: on the one hand the systematically ordered text, of which Peter Kolb's *Capvt Bona Spei Hodiernvm* of 1719 is an example; on the other hand an account in which the chronology of the journey is largely maintained.[46] In this case the chronology also had an epistemological function. According to Sparrman, following the pattern of the journal was the "most natural" method. As the reader has an idea of the time, circumstances and the sequence in which the events occurred, he also has a better understanding of what is described. But the main advantage is that the reader can distinguish between what the author observed and what he is asserting on the authority of others.[47] It was also possible to deal with certain subjects systematically, independently of the chronology, by way of a digression. In this way information that was collected in different places in the course of the journey was arranged together for the reader's convenience. This often happened when indigenous peoples were described. The published travel accounts from the last quarter of the eighteenth century all followed the chronological pattern.

Towards the end of the eighteenth century, attention to the presentation of the published travel account became more and more important.[48] Sparrman, Le Vaillant, Barrow and Lichtenstein attempted to please the reader by employing a well-polished style. Owing to the influence of published travel literature, this attention to style can be discerned even in the unpublished travel reports of Batavian officials (chapter 8). Le Vaillant and his editors made the greatest efforts to increase reading pleasure by inserting anecdotes—which the serious John Barrow felt were rather tasteless. For Le Vaillant, the traveller is no longer a neutral observer; through his conduct—and that of his pets and his

[45] Chapter 6.

[46] Travel accounts of the seventeenth and eighteenth centuries were arranged either systematically or chronologically; see Bourguet (1996, pp. 328–331); Rubiés (2000, pp. 25–26); Shapiro (2000, p. 70).

[47] Sparrman (1975, p. 36).

[48] Batten (1978).

personnel—he demands the reader's attention. He regularly dishes up anecdotes about his baboon Kees, which was still a "virgin"; tells how he acted as peacemaker during his journey to the north-west and about his fascination with a Gonaqua girl he called Narina ('flower'). Georg Forster, who was no less serious, took little notice of this. In a long review he effusively praises Le Vaillant because he succeeded so well in evoking a complete picture of the regions he had crossed. In comparison, the products of "tasteless industriousness" were nothing more than random assemblies.[49] In the eyes of some readers, literary efforts could therefore enhance the representation.

Providing a total picture of the foreign world was the main function of the travel account. Despite the great popularity of natural history, the scientific authors of travel accounts at the end of the eighteenth century were therefore careful not to expand too much about botanical and zoological subjects, although Thunberg and Sparrman in particular, being pupils of Linnaeus, sometimes could not refrain from inserting fairly long descriptions of plants and animals. Nevertheless, this did not turn the travel account into a botanical or zoological treatise. Sparrman states explicitly that botanists should not expect to find a catalogue of plants in a travel account. They would have to be satisfied with information that was given in passing in order to get an idea of the vegetation.[50] The author is careful not to bore the reader with too much specialised knowledge of natural history; this was published in scientific journals or in a botanical catalogue such as Thunberg's *Flora capensis* (1807). Although at least Thunberg and Sparrman must be called natural historians by profession, their travel accounts cannot be classified as belonging to natural history in the narrow sense, because descriptions of nature ultimately form a subordinate part of an attempt to produce an encyclopaedic total impression of the foreign world.

The attempt to present a total impression was completed by adding a map and figures.[51] This helped to put the reader in the same place

---

[49] Forster (1992, pp. 225–226).

[50] Sparrman (1975, p. 66).

[51] Most of the travel accounts discussed here contained a map: Sparrman (1974/5), Paterson, Le Vaillant, Barrow, Lichtenstein. Illustrations are used by Sparrman, Le Vaillant and Lichtenstein. The ethnographic monograph by Alberti (1810) contains a map and figures. Godlewska (1995, pp. 11) makes the following statement about the interaction of text, maps and illustrations in the *Description de l'Egypte* (Description of Egypt): "The sketch offers a level of completeness and a strength of image and impression which is not found in the text but which flickers in the map. There is a

as the traveller, "where he is able to imagine and picture everything exactly in terms of his own sensibility".[52] Maps and illustrations offered an additional visual perspective of the region described in the text. The travel account with figures and maps had to produce a portable miniature of the foreign world.[53]

The map was not only intended to give an overview of the topography of the foreign region, but, according to Lichtenstein, served in the first place to allow the travel experiences in the text to be located. For this reason the route followed was indicated on most of the maps and the places and (sometimes) observations were indicated: "As our map serves mainly to illustrate the description of the route followed on the journey, all places mentioned in the book have been carefully indicated on the map."[54] The number of notes on the map often increased in regions that had not yet been frequently described (fig. 46).

In addition, travellers collected specimens and made drawings of plants, animals, people, utensils and landscapes. They drew these themselves (Sparrman, Le Vaillant, Barrow) or used drawings made by subordinates (Gordon) or acquaintances (Paterson, Lichtenstein, Alberti). The figures had to give a representative picture of the distant reality. Especially around 1800, trueness to nature became very important. If the draughtsman had made too much use of his fantasy, his sketches were useless as illustrations. Lichtenstein, in any case, was so conscientious that he explained deviations from reality in the annexures. For example, a wagon appears to be moving in the wrong direction in a drawing when the drawing is compared with the text, while elsewhere the original drawings are collated in the book illustration. On the whole, however, there is a close relation between illustrations and texts. Illustrations and texts supplement one another.[55] The problem of depicting the foreign world will be separately dealt with in the chapter on ethnographic drawings in the Gordon Atlas (chapter 5).

---

precision in the text and the map which the sketches strive for but fail to achieve. In the *Description* the genres reach for each other and almost merge." She also mentions a "unity of text, image and map".

52 Forster (1784).

53 Latour (1987, p. 227) refers to "immutable mobiles" in this sense.

54 Lichtenstein (1811, p. 681).

55 Lichtenstein (1811, pp. 683–685; 1812, pp. 635–640). In general: Stafford (1984, p. 59).

Cape ethnography

The objective of many travel accounts was to give a total impression, a cross-section of the foreign world. Despite the huge popularity of natural history and despite the fact that for example Sparrman, Thunberg and Le Vaillant were themselves naturalists, their travel accounts were not limited to a description of the "three kingdoms of nature". All travel reports also have something to say about the culture of the indigenous population. The extent of the ethnographic description can vary from incidental remarks to (in exceptional cases) a whole book about an indigenous population group, as in the case of *De Kaffers aan de Zuidkust van Afrika* ('The Kaffirs on the South Coast of Africa'; 1810) by the Batavian army officer Lodewijk or Ludwig Alberti (1768–1812; chapter 9). Because of the ubiquity of ethnographic descriptions in travel accounts, I will emphasise this aspect in many chapters.

As most of the interest was focused on the unassimilated Khoikhoi or "Hottentots" (and later also on the "Kaffirs" or "Kaffers"—Xhosas) which at the end of the eighteenth century could only be found in the farthest corners of the colony and especially beyond them, extensive descriptions of indigenous inhabitants are only found in the works of travellers who journeyed farther into the interior and stayed there for longer.[56] Anyone who did not penetrate far into the interior and who, like Thunberg and Sparrman, did not understand the language of the indigenous inhabitants or Dutch was unable to do much more than produce descriptions of the external appearance of the small groups of Khoikhoi he encountered within the borders of the colony. Le Vaillant, Barrow, Lichtenstein and Alberti, however, who had more time at their disposal, had interpreters or spoke Dutch (Gordon, Le Vaillant, Lichtenstein and Alberti), provide extensive descriptions of the peoples they visited in the interior.

Concern about the limited duration of the contact between travellers and the indigenous population and about the unfamiliarity with the indigenous languages was already expressed at the time in one of the first anthropological treatises, the *Considérations sur les methodes à suivre dans l'observation des Peuples Sauvages* ('Considerations Concerning the Methods to be Followed when Observing Savage Peoples') by Joseph-Marie Degérando (1772–1842) in 1800. According to Degérando one

[56] Thunberg (1986, p. 79).

had to live among the foreign population for a longer time in order to get to know it properly. The deficient descriptions were, according to Degérando, caused by the inability to communicate with the indigenous population. As the travellers did not speak the language, they could not get to know anything about the customs and the systems of thought of foreign peoples (see chapter 9). Degérando had observed this correctly, because travellers without time and the necessary linguistic skills usually had to limit their description to the external characteristics of the few 'wild Hottentots' they came across.

The inhabitants of Southern Africa are usually not described in the chronological framework of most of the travel accounts, but systematically in passages in which appearance and customs are captured. These systematic descriptions arise from the deliberate concentration of observations made in various places and at various times. For the reader's convenience they are grouped together in excursions written in the "ethnographic present".[57] Such an excursion had to describe the customs of a people in an unspecified present. The suggestion that emanates from descriptions of this kind is that the customs are immutable characteristics of the people in question. The systematics were obtained by giving the description a topical structure that often depended on the publications of predecessors. As a reader one could thus quickly gain an impression of the appearance and customs of a certain population group, which would have been more difficult if the observations had been dispersed throughout the travel account. This concentration of observations was therefore more a rhetorical means to present information clearly than an effort to create an impression of objectivity. Instead, it was attempted to create an impression of objectivity of the observation by criticising predecessors. By pointing out errors in the works of earlier authors, the author of a travel account proved his own reliability and established his authority.[58] Unfortunately, this authority was of a temporary nature, because subsequent travellers would reject the authority of their predecessors in favour of their own claims. Thus Le Vaillant criticised Kolb, Sparrman and Thunberg; Barrow regularly rejected the findings of all the above-mentioned travellers and Lichtenstein, the last in the row, corrected those of Barrow to

[57] Van Maanen (1988, p. 68).

[58] This was an old procedure in the scientific literature and was already applied by Strabo (64 B.C.–21 A.D.). (Godlewska, 1999, p. 95).

boot. Inevitably, this cast doubt on the claims of veracity of *all* travel accounts.

The population groups living in the region visited were not only described, but also explicitly evaluated. Differences between population groups were generally not yet reduced mainly to differences in physical characteristics, as became customary after the emergence of nineteenth-century scientific racism. 'Race' is a concept that was only developed at the end of the eighteenth century by a few scientists such as Immanuel Kant (1724–1804) and Johann Friedrich Blumenbach (1752–1840); however, it did not acquire a fixed meaning and wide distribution in anthropological discourse before 1800. Until the end of the eighteenth century different groups of people were not generally called human races, but 'varieties'.[59] The European sense of superiority was not as strongly developed either. In the eighteenth century, Asian civilisations—particularly those of China and Japan—were regarded as being equal and in many cases even superior to those of Europe.[60] Many scholars, such as Johann Gottfried Herder (1744–1803), held that the most beautiful people did not live in Europe, but in the region between the Caucasus and India.[61] Eighteenth-century travellers did of course have prejudices against people who looked different from themselves, but the systematics were still lacking, so that these prejudices cannot be called racism yet, since "racism is related to racial prejudice like dogma is to superstition. Racial prejudice is fairly fragmented and full of contradictions [...]. Racism is relatively systematic and internally consistent [...]."[62]

For evaluating local population groups, the eighteenth-century travellers in South Africa basically employed two criteria which sometimes produced contradictory evaluations. First of all, all travellers

[59] Wheeler (2002, p. 31). In the literature on race and racism it is generally accepted that the dominance of the concept of race in the anthropological discourse is a nineteenth-century phenomenon; see inter alia: Stepan (1982), Blanckaert (1988, pp. 18–55), Douglas (1999, pp. 82, 91, 2003, pp. 3–27), Wheeler (2000) and Bindman (2002). Robert Young (1995, p. 7) summarises the eighteenth-century opinion about race as follows: "The dominant view at that time was that the idea of humans being of different species, and therefore of different origins, conflicted with the Biblical account."

[60] Osterhammel (1998), Fuchs (2006).

[61] Bindman (2002, pp. 25, 166).

[62] Fryer (1984, p. 134). The perceptions of 'others' were not firmly established yet. The Chinese, for example, only turned 'yellow' to European observers in the course of the eighteenth century. Until then, they were white to most Europeans (Demel, 1992).

used a vague civilisation criterion that became slightly explicit only when Lichtenstein compared "Hottentots" with "Kaffirs". The Kaffirs (Xhosas), according to Lichtenstein, are a better than "halbgesittetes Volk" (half-mannered people) and the "Beetjuanen" (Tswana) are "mehr als halbgebildet" (more than semi-civilised).[63] In comparison, the "Hottentots" are on a lower level in the hierarchy for very diverse reasons: lack of physical strength and external beauty, poverty of language and mind, the absence of laws and because property is partly unknown to them. They are as different from the Kaffirs as "der Muselmann von dem Britten" (the Moslem from the Briton).[64] Lichtenstein also regards order and refinement of language as a criterion of the "Kulturgrad" (cultural level) of a population group. In the language of the Xhosas, for example, he misses the order and regularity of declension: the "innere Konsequenz der Sprache" (internal consistency of the language) is absent.[65] The language of savages has no "Geist" (is without a soul) and is intended for communication only.[66] Incidentally, this also applies to the language of the colonists, which is poor in distinguishing concepts.[67] For Barrow, physiognomic characteristics put the Xhosas on a higher level in the human hierarchy.[68] He even thought that on the basis of the external appearance they merited a position next to "the first Europeans".[69] Moreover, Xhosa women could be called attractive if one was able to overcome the colour prejudice.[70] Barrow and Lichtenstein used ethnocentric criteria. The different South African population groups were assigned a place in a civilisation hierarchy in which the culture and body of north-western Europeans constituted the model and the degree of deviation from this model determined the position in the hierarchy.

The moral criterion that is used, however, partly contradicts the civilisation hierarchy. For example, although the colonists are higher up the (technical) ladder of civilisation, they are to most travellers also morally the most repugnant group. For Barrow and Lichtenstein, the Bantu-speaking "Kaffirs" (Xhosas) and "Beetjuanen" (Tswanas) are at

[63] Lichtenstein (1811, p. 395; 1812, p. 511).
[64] Lichtenstein (1811, p. 395).
[65] Lichtenstein (1811, pp. 672, 664, 635).
[66] Lichtenstein (1811, p. 636).
[67] Lichtenstein (1811, p. 214).
[68] Barrow (1801, p. 206).
[69] Barrow (1801, p. 206).
[70] Barrow (1801, p. 168).

the top of the moral hierarchy. Khoikhoi groups, such as the Gonaqua, can possess all sorts of qualities that make them superior to the colonists and the "Bastaards" who are descended from the colonists and Khoikhoi. In this evaluation, the Bastaards are mainly the heirs of the European colonists. In Le Vaillant's view, they share with them dynamic action (which determines their position in the civilisation hierarchy) and moral repugnancy. Le Vaillant sees in this group an ominous sign for the future. On the other hand, the Bastaards, which were the result of relationships between European colonists and indigenous women, retained the bad properties of their white ancestors. The combination of Khoikhoi and Xhosa produced much better results, but Le Vaillant nevertheless foresaw that the noble savages in the expanding colony would ultimately be replaced by the "Bastard-whites".[71]

A combination of these two criteria—level of civilisation and moral qualities—arouses in the authors a hierarchy of sympathy and repulsion. In this regard the Xhosas and, for Lichtenstein, the "Beetjuanen" are the darlings of the European travellers. Together they form a group that Lichtenstein would rather call civilised than uncivilised and that Barrow wants to put side by side with the Europeans on the basis of anatomical and physiognomic similarities. Gordon was even a friend of the Xhosa chief Coba (fig. 18). However, not all scientific travellers have much to say about the Xhosas, because some travellers did not come into sufficient contact with them. Thunberg, for example, knows so little about the Xhosas that he cannot distinguish them from the Khoikhoi.

Lichtenstein and Barrow regard the Xhosas as more than half civilised not only because of their appearance. Lichtenstein's "Beetjuanen" are also people who like cleanliness and love their tools.[72] Everything shows that they know laws and live a modest life. The bad name the Xhosas have in some circles is the work of the colonists. If one compares their conduct towards people who have been shipwrecked with that of the colonists, one finds that it is the latter who are the real "savages".[73] It was about these Kaffirs or Xhosas that the first ethnographic monograph with a South African population group as its subject appeared, namely *De Kaffers aan de Zuidkust van Afrika* ('The Kaffirs in the South Coast of Africa') by Lodewijk Alberti in 1810 (chapter 9).

[71] Le Vaillant (1790, pp. 262–266).
[72] Lichtenstein (1812, pp. 504, 513).
[73] Barrow (1801, p. 196).

The evaluation of the Hottentots (Khoikhoi or Khoekhoen) is more equivocal. Early in the eighteenth century Kolb did his best to improve their image after visitors to the Cape had depicted them as stinking, stuttering and repugnant creatures for two centuries. Kolb's favourable representation of the Khoikhoi was used by Rousseau in his *Discours sur l'inégalité* ('Discourse on Inequality'; 1755), inter alia to lay an empirical basis for his representation of the noble savage (chapter 2). In a certain sense, Rousseau's noble savage returns to South Africa in Le Vaillant's description of the Gonaqua, who are of mixed Xhosa and Khoikhoi descent.[74] To Le Vaillant they are the ideal type of human being in his natural state. In his description he gives not only the usual summary of their products, manners and customs and their external characteristics, but—to the annoyance of some of his readers—even dramatises his representation with a romance between himself and a Gonaqua girl he calls Narina ('flower' in the Gonaqua language). This romance is a suitable means of reversing the negative image of the Hottentot that had prevailed since the sixteenth century, a reversal that was still necessary despite Kolb's efforts. Whereas the Hottentot woman was generally held to be particularly unattractive, Narina made an overwhelming impression of beauty (chapter 6).

Le Vaillant's enthusiastic representation of the Khoikhoi as embodied in the Gonaqua is not common among the scientific travellers. The others also lack the primitivistic ideal of the free human being in a primitive society. To Thunberg the Hottentots are (in the English translation) "the most wretched of the human race", "scarcely differing from the wild beasts".[75] To Barrow they are low down on the "scale of humanity" and Lichtenstein finds them a lot of incorrigible savages.[76] Nevertheless, Barrow does realise that the negative representation may be one-sided, and Thunberg finds that in any case they were treated unfairly.[77]

There is even less inclination to express a favourable view of the "Bushmen" (San). Le Vaillant does express himself favourably about the "Houzouanas" in the north-west of South Africa, but not about other San, whom he regards as a pack of absconded slaves and other crooks living as robbers. To Barrow they are "savages", partly because

[74] Sparrman (1977, p. 15).
[75] Thunberg (1986, pp. 319, lvii).
[76] Barrow (1801, p. 151); Lichtenstein (1811, p. 253).
[77] Barrow (1801, p. 156); Thunberg (1986, p. 289).

of external characteristics: "Nature seems to have studied to make this pygmy race disgusting".[78] Lichtenstein also finds them terribly ugly. To him they are animals, without any idea of property, without close community ties or fixed abode, using a language that has but few words and sounds. Consequently they cannot reach a higher rung of the cultural ladder on their own. Their only hope is the establishment of a reformatory on Robben Island, where they could be forced to learn to work.[79] Nevertheless, it is also realised that the Bushmen have been treated badly and unfairly.[80] Barrow in particular shows this in his description of the behaviour of a punitive commando consisting of Europeans and Bastaards (chapter 7).[81]

This brings us to the group that comes off worst among scientific travellers in many regards: the Dutch colonists, sometimes already called "Boers" at this time.[82] In the primitivistic perspective of Le Vaillant a critical view is to be expected. After all, in his view peoples such as the Gonaqua are noble savages who risk losing their blissful life to the advancing 'sick' civilisation.[83] Barrow viewed the colonists through the glasses of the English coloniser with a high humanitarian self-esteem. He regarded the degenerated state of the Dutch colonists in South Africa as proof of the superiority of British colonialism. The image of the colonists was strongly influenced by the travellers' ideological preferences. I will discuss this in more detail in the chapters on Le Vaillant and Barrow.

Approaches

In studying the South African travel accounts, the existing critical literature broadly follows two approaches: on the one hand, in the older studies, a positivistic reconstruction of the routes followed by the expeditions, and on the other, in recent literature, a postcolonial accusation against colonial representations.

---

[78] Barrow (1801, pp. 234, 278).

[79] Lichtenstein (1811, p. 89; 1812, pp. 76, 81–82, 313, 318, 343, 366).

[80] Sparrman (1977, p. 112).

[81] Barrow (1801, pp. 269–272).

[82] The 'Dutch colonists' in South Africa were descendants of Dutch, German, French, Khoikhoi and slave ancestors and mostly spoke vernacular Dutch.

[83] Le Vaillant (1790, pp. 257–258). We will see in chapter 6 that Le Vaillant is not entirely consistent in his primitivistic attitude.

The first approach is found from the start of the twentieth century until the 1970s in studies and text editions by E.C. Godée Molsbergen, E.E. Mossop, V.S. Forbes and W.H.J. Punt, to name the most important.[84] They worked within the programme of colonial (even colonialist) history in which journeys of exploration were seen as the first attempts to open up the wild interior to the creative forces of European civilisation. These historians focused especially on travel routes in their research. By correlating information provided by travel journals with modern maps, following the hypothetical routes by car or on foot and comparing the landscape with the descriptions, they attempted to reconstruct the routes followed by the discoverers (figure 6). This approach is an extreme case of Ranckean "wie es eigentlich gewesen [ist]", actually rather *wo* "es eigentlich gewesen" (*where* it actually was); often with the aim to be accurate to within a few metres, sometimes with highly disputable results.[85] In the background, a colonial ideology plays a part in which exploration is a prelude to the establishment of a white-dominated South African state in the twentieth century. The fact that the footsteps of the explorers heralded the coming of European civilisation is what made them so admirable in the view of these historians. One quotation from *Pioneer Travellers of South Africa* by Vernon Forbes (1965) suffices to illustrate this attitude towards eighteenth-century South African travel accounts. Forbes concludes his reconstruction of the route of a VOC expedition in 1752 led by ensign Beutler that visited the Eastern Cape as follows:

> On or near the route pioneered 200 years ago by that long waggon train of sturdy adventurers, their successors inherit the land and peopled its towns—Port Elizabeth, Grahamstown, King William's Town, East London, Butterworth and Cradock. The development of a region must be preceded by foreknowledge of its potentialities. Amongst the many who laboured to this end, Beutler and his men played a notable initiatory part.[86]

---

[84] This refers inter alia to the following works: *Reizen in Zuid-Afrika in de Hollandse tijd* (Godée Molsbergen, 1916–1932); *Journals of the expeditions of the Honourable Ensign Olof Bergh (1682 and 1683) and the Ensign Isaq Schrijver* (1689) (Mossop, 1931); *Old Cape highways* (Mossop, s.a.); *Pioneer travellers of South Africa: a geographical commentary upon routes, records, observations and opinions of travellers at the Cape, 1750–1800* (Forbes, 1965); *Die eerste Europeane in die Nasionale Krugerwildtuin, 1725; The first Europeans in the Kruger National Park, 1725* (Punt, 1976). These authors have produced more publications on this subject, but I mention only the most important here.

[85] An example is given in chapter 3.

[86] Forbes, 1965, p. 24.

The quotation is a colonial topos: it expresses admiration for the "sturdy adventurers" who play their part of ground-breakers for Western civilisation and progress, seen from the perspective of their twentieth-century, white successors in the Eastern Cape. The description virtually demands a statue or a long relief.

The new approach is in many ways the ideological mirror image of this approach and is nurtured by indignation about apartheid and colonialism. As regards research of the eighteenth-century representations of South Africa, the tone was set especially by the publications of the Canadian-American Hispanist Mary Louise Pratt, first in an essay on John Barrow, "Mr Barrow in the land of the Bushmen" and later in the first chapters of her book *Imperial Eyes* in which, besides John Barrow, she also involves Peter Kolb, Anders Sparrman and William Paterson in her arguments.[87]

Pratt sees travel texts from the second half of the eighteenth century as attempts to intellectually annex non-European territory and to pave the way for colonial expansion. As part of the world-encompassing project of natural history, travellers tried since the last quarter of the eighteenth century to describe South African nature in terms of Western intellectual frameworks. Under the influence of natural history, these travellers are said to have become instruments of the deployment of colonial power. Because of their focus on natural phenomena, they ignored the presence of the indigenous population in their representations or degraded the population to natural phenomena. This created the impression that there was an empty land ready for colonial expansion.

Whereas the old historical school of Forbes and his fellows used South African travel texts for hyper-accurate reconstructions of the routes which expeditions followed in order to open up the interior for Western Civilisation, Pratt's approach is characterised by a postcolonial *tiersmondisme*. She uses the travel texts to "illustrate" the complicity of Western science in colonial expansion in the eighteenth century by resorting to fragments from a few travel texts without taking the complexity of the texts as a whole into account.[88] Her pronouncements on the texts in question are therefore highly debatable.[89] The ideological

---

[87] Pratt (1986, 1992).

[88] Pratt (1992, p. 39).

[89] Compare the following criticism: Beinart (1998); Huigen (1998); Guelke & Guelke (2004). See also chapters 7 and 10.

effect is the opposite of Forbes's approach, but in both cases the texts themselves are overlooked: their structure, ideological ambiguity and the complex interconnectedness with intellectual contexts.

Pratt's approach is also rooted in the conviction that writers are subject to discursive formations, such as natural history or colonial discourse, which determine how they represent the South African reality. In the following chapters I want to show, however, that a broad spectrum of possibilities existed between tight discursive restriction and agency. Some authors could claim a high degree of agency and used it. If this had not been the case, Peter Kolb would not have been able to change the direction of the way in which South Africa was being represented and (I am taking a leap here) Mary Louise Pratt's postcolonial criticism would have been impossible.[90] VOC travellers, on the other hand, were subject to explicit restrictions in their reporting. They had to produce answers to specific questions their principals had put. Their travel accounts consequently suffer from uniformity and a dearth of information compared with the work of scientific travellers. The scientific travellers in South Africa had more latitude because their behaviour was not regulated by centres of calculation. Moreover, their agency was a consequence of their view of science. The empiricist paradigm within which all scientific travellers operated sharpened their observation of the foreign world. They were looking for fragments of experiential knowledge that could contradict accepted representations.[91] Furthermore, the scientific traveller needed these new facts to establish his own authority vis à vis his predecessors. The representations of

[90] Dieter Freundlieb (1995, p. 175) analyses discursive determinism, which is also dominant in the theory of colonial discourse, in an essay on Foucault's archaeological method. He shows that it is based on a sophism. If the discourses analysed by Foucault are dictated by arbitrary rules and say nothing about reality, as Foucault argues, then the same applies to Foucault's archaeological discourse theory, Freundlieb says. According to Freundlieb, human agency is an indispensable element in the history of science: "[...] we cannot begin to understand the history of the production of knowledge unless we recognize the role or the ability of human beings to engage in individual and collective forms of solving cognitive problems. This form of human agency is crucial, and while Foucault tries very hard to eliminate it from his archaeological project, it keeps resurfacing in his vocabulary and through his habit of metaphorically attributing agency to systems of rules which cannot perform acts but rely, to a large extent, on human beings who can *apply* and, if necessary, modify or abandon those rules."

[91] Foucault describes the "observing gaze" as follows: "The observing gaze manifests its virtues only in a double silence: the relative silence of theories, imaginings, and whatever serves as an obstacle to the sensible immediate; and the absolute silence of all language that is anterior to that of the visible" (Foucault, 2003, p. 132).

the foreign world were therefore the product of the constant diligence with which travellers sought to present 'true' facts and to correct their predecessors. In collecting new knowledge, at least as much emphasis was put on holding a different opinion than the predecessor as on adducing supplements to the existing discourse. Proceeding from the existing discourse, each traveller constructed his own representation of the South African world.[92] As a result, the representations of scientific travellers have too much of an individual profile to allow them to be subordinated to an amorphous and supra-individual discourse in the analysis.

With their representations, the travellers attempted to pursue their own goals, but of course they did this against the background of what others had already contended and also of a shared empiricist epistemology. A dialectic existed between individual intention on the one hand and already existing representations and cognitive group characteristics on the other (chapter 10). Even Batavian representations, which mostly came about with administrative objectives, sometimes display individual characteristics. However, the agency always remains relative. With this in mind, I have therefore endeavoured in the following chapters to put the spotlight specifically on the uniqueness of the representations of scientific travellers. VOC representations, by contrast, tend towards uniformity, and in that respect they offer a good counterfoil that sets off the more individual profile of the other travel accounts better.

Many travellers who are discussed in the following chapters do not satisfy the image currently prevailing in the postcolonial theory, namely that of the narrow-minded European who already knew before his departure what was wrong with the natives.[93] On the whole, the travellers of the eighteenth century took pains to understand foreign cultures, were inquisitive and had the latest knowledge at their disposal. Their achievements were sometimes considerable. Within a few

---

[92] The sociologist of science Andrew Pickering (1999, p. 375) says the following about the agency of scientists, with whom scientific travellers display many similarities: "I find that I cannot understand scientific practice without reference to the intentions of scientists [...]. The key remark, for me, is that we humans live in time in a particular way. We construct goals that refer to presently nonexistent future states and then seek to bring them about. I can see no reason to suppose that DNA double helices or televisions organize their existence thus—why should they?"

[93] This image of the European travellers and their representations of the non-Western world has become widely accepted in studies of colonial discourse following the publication of Edward Said's *Orientalism* in 1978.

years, Peter Kolb wrote a colossal encyclopaedia in which all available knowledge on South Africa had been collected and in which he did his best to restore the damaged image of the Khoikhoi. John Barrow was an autodidact, but succeeded in determining the European view of South Africa and China until deep into the nineteenth century. Their achievements imbued these travellers with a professional pride that was expressed in criticism of armchair scholars. For a number of scientific travellers (Le Vaillant, Gordon and Barrow) the centre of the production of knowledge did not lie in European scientific institutions, but in their own ox-wagon, on the outskirts of the known world. Their self-awareness befitted, in a certain sense, the scientific paradigms of their time. If direct observation of phenomena was indeed the main issue, then the field worker might well be more important than the person who catalogued the observations in his isolated study. I will return to this issue in the concluding chapter.

The sequence of the chapters between the introduction and the concluding chapter reflects the chronology according to the origin of the representations described in them. At the same time, the chapters also follow a thematic order based on three objectives that played a part in the drafting of representations: the VOC's endeavours to investigate the economic potential of the interior, the efforts of scientifically interested individuals to gather knowledge about South Africa and the Batavian government's wish to get an administrative hold on the colony and its frontiers. This is a provisional, schematic classification, because when we take note of the networks of which the producers of representations and the representations themselves formed part, matters are rather more complicated in practice. This problem area will also be dealt with in the concluding chapter.

In chapter 2 on Peter Kolb's *Capvt Bonae Spei Hodiernvm* (The Current State of the Cape of Good Hope) of 1719 the new scientific approach to the Cape is introduced. Chapters 3 and 4, dealing with representations by travellers in the service of the VOC, chronologically (1719–1778) follow the chapter on Kolb, but in terms of content they connect to the seventeenth-century VOC discourse on South Africa. They are in contrast to chapters 5, 6 and 7 on scientific representations of South Africa by Robert Jacob Gordon, François le Vaillant and John Barrow, travellers from the last quarter of the eighteenth century who reacted critically to Kolb. Lastly, chapters 8 and 9 are devoted to representations which owe their existence to the efforts made to restore Dutch rule over the Cape

Colony during the Batavian period (1803–1806). These representations were largely produced with administrative objectives.

Ethnonyms

In conclusion, a remark about the ethnonyms used here and in the following chapters. In recent decades, terms that may be experienced as discriminatory are preferably avoided in scholarly writing about South Africa. For example, 'Khoikhoi' is preferred to 'Hottentots', because the latter concept has a discriminatory meaning, and in addition there are good reasons to assume that these people referred to themselves as Khoikhoi or Khoekhoen (or a variant of these terms). However, in research of eighteenth-century European representations the use of terms such as Hottentots (for Khoikhoi), 'Kaffirs' (used in the eighteenth century for Xhosas, now a derogatory term for 'blacks', 'Africans' in South Africa) and 'Bushmen' (San) is unavoidable because these were the common concepts in European representations at the time. In any case, the currently used terms are not unproblematic either. The term 'San', introduced with the best intentions to get rid of the term 'Bushmen', is actually discriminating; in the language of the Khoikhoi it means something like 'robbers' or 'stock thieves'. The South African anthropologist Schapera is responsible for the current popularity of the term 'Khoisan' as the umbrella term for Khoikhoi and San. However, this concept suggests a homogeneity that never existed. As for the term 'blacks', it is totally unclear who is referred to: all people in South Africa who are not 'white' or only some members of this group.

When analysing representations from another period, it is desirable to use the designations that were current at the time. I am concerned primarily with the *representations* of the indigenous population of South Africa constructed by Europeans, and the European terms that were current at the time are part of this. But I realise that the repeated use of the old ethnonyms may be irritating to readers who are aware of the nomenclature issues, and therefore I frequently switch to the terms in current use where this cannot cause confusion. In order not to clutter the typography, I have used quotation marks for the controversial concepts only sparingly.

CHAPTER TWO

PETER KOLB'S DEFENCE OF THE "HOTTENTOTS" (1719)

In the night of 26 April 1712, in Stellenbosch, Peter Kolb (1675–1726; fig. 7) suddenly lost his vision.[1] When he wanted to read, it felt as if "thick sand" had been thrown into his eyes; it made no difference how large or how clearly the words were written. As he was secretary of the VOC, this was of course a great nuisance. He tried everything to cure it: a string of horse hair through his neck, an artificial abscess to allow the bad humours to escape and swallowing concoctions containing Spanish fly. Of all these cures Spanish fly was the least effective.[2] A visit to a competent doctor in Baden-Baden in 1713 fortunately restored his vision, and a few years later he wrote the most comprehensive eighteenth-century book on the Cape, *Capvt Bonae Spei Hodiernvm* (The Present State of the Cape of Good Hope; fig. 8.) This book of 900 folios, with illustrations and index, appeared in Nuremberg in 1719 and was followed by translations and editions in Dutch (1727), English (1731, partly in 1738), French (1741, partly in 1742 and 1743) and then again in German (1745).[3] From an ethnographic point of view, Kolb's book is the most important work on the Cape to appear between 1500 and 1800. The book ended a period in the European representation of southern Africa during which its image was determined by travellers of the sixteenth and seventeenth century who dedicated a few pages to their short stay at the Cape in the accounts of their travels between Europe and Asia. Besides remarks on the geography and the colonial settlement (after 1652), this period produced stereotypical and extremely depreciatory descriptions of the indigenous population, the Khoikhoi or Hottentots.[4] Typical of this type of representation of the Hottentots is the final assessment arrived at by Wouter Schouten (1638–1704):

---

[1] Kolb's name is given as "Kolbe" in the Dutch translation of his work. Below I use his name as it is spelt in Kolb 1719.

[2] Kolb (1719, "Dedicatio", p. 838).

[3] See Annex 1 for the main eighteenth century editions of Kolb's work.

[4] As this analysis is aimed at the European representation of the Khoikhoi, I will also sometimes use the contemporary European terminology.

> It is deplorable that among mankind there are such people (as we have now shown) who, although also descended from our first father *Adam*, display nevertheless so little humanity that they truly are more like senseless animals than reasonable people, living a miserable and wretched life on the face of the Earth, having no knowledge of God nor of the road to salvation. Miserable people, how lamentable is your pitiful state! And oh Christians, how blessed is ours![5]

The badness of the Hottentots was established a priori; the author merely made a selection from the existing reasons. Kolb's achievement was that he attempted to turn this negative opinion around, and furthermore that he attempted to provide full information. In this way he transformed the discourse on the Khoikhoi. What his predecessors disposed of in few words he dealt with exhaustively, from different perspectives and comparing the existing literature with his own observations. Kolb transformed the Khoikhoi from a stereotyped subject in a travel account that mentioned South Africa merely in passing to a complex subject of scholarly research in a book that dealt exclusively with the Cape. At the same time Kolb's book laid the foundation for a scientific interest that treated South Africa as an independent theme and for an anthropology that viewed the Khoikhoi as people with a culture that was not automatically disposed of as inferior. He transformed the old discourse about the Hottentots into research questions. He examined the tenability of all clichés and then usually rejected them. After Kolb this approach gained more ground, although towards the end of the eighteenth century the focus on the Khoikhoi decreased—on the one hand because the Xhosas appeared on the scene and on the other because unassimilated Khoikhoi became increasingly difficult to find, and then only farther away from Cape Town. Nevertheless, when the Khoikhoi and the San (who are related to them) were being described, Kolb's work was still referred to in such descriptions years later. Decades after the publication of *Capvt Bona Spei Hodiernvm* he was still quoted by Gordon, Le Vaillant, Barrow and Lichtenstein.

[5] "Beklaeghelijck is 't / dat onder het Menschdom sulcke Menschen (gelijck wy nu hebben getoont) gevonden worden / die / schoon mede van onsen eersten Vader *Adam* afkomstigh / nochtans soo eynigh Menschelijkheyts vertoonen / datse warelijck meer het onvernuftige Vee / dan redelijcke Menschen komen gelijck te zijn / levende op den Aerdbodem een soo rampsaligh en elendigh leven / hebbende gantsch geen kennisse Godts / noch die ter saligheyt leydt. Elendigh Volck / hoe beklaeghelijck is uwe erbaarmelijcke stant! en ô Christenen, hoe gelucksaligh de onse!" (Schouten 1676, p. 184).

In historiography, Kolb's significance is largely overlooked. His work is, however, a gratefully acknowledged treasure house for historical anthropologists. Books on the ethnography of the Khoikhoi, such as those of Schapera and Barnard, rely to some extent on the partial English translations of Kolb's work.[6] In recent studies of the representation of the indigenous population, Kolb is also mentioned, and he is—especially by many English-speaking historians—unjustly counted among the European authors who presented a caricature of the indigenous population of southern Africa. This view of Kolb is, however, based on careless reading of the partial English translation of Kolb's magnum opus.[7] The difference between the original German text and English and French translations of Kolb is already evident from the difference in weight. Whereas the original Kolb is a book in folio format weighing 3 kg (6.6 lbs), the English and French translations consist of two and three little octavo volumes weighing not more than 1 kg (2.2 lbs, see annex 1) respectively. Only the Dutch translation is comparable to the German original in most respects; moreover, it has more attractive illustrations.

This chapter examines the way in which Kolb attempted to construe the new image of the Khoikhoi. In the existing literature this aspect has remained underexposed.[8] In my view, the central issue of Kolb's impressive book is that he takes a stand against the customary views of

[6] Schapera (1930), Barnard (1992).

[7] Landau (1996, pp. 129–142); Chidester (1996, pp. 47–67); Hall (2000, pp. 118–120, 127); Hudson (2004, pp. 318–320). Although Mary Louise Pratt, in *Imperial Eyes* (1992, pp. 41–49) praises Kolb as one who enters into dialogue with the indigenous population, she does this on the wrong grounds and uses only the first part of the incomplete English translation of Kolb's work, which left her with the incorrect impression that Kolb took no interest in Cape nature. This misconception plays a major role in the rest of *Imperial Eyes*—see Huigen (1998, pp. 75–76). Pratt's view is followed by Merians (2001, pp. 151–155), who also used only the English translation. It is remarkable that none of the researchers mentioned in this note even deemed it necessary to consult the full Kolb text when they set out to characterise his work. It is not customary to approach canonised literature with such a degree of carelessness.

[8] In the English-language historiography Peter Carstens, the editor of the modern reprint of the English translation (Kolb, 1968, pp. v–xi), Pratt (1992) and Merians (2001) adopt a more favourable opinion of Kolb. In German studies it is generally assumed that Kolb presents a more favourable image of the Khoikhoi, but how he does this has never been thoroughly researched (Dos Santos Lopes, 1992, pp. 86–89; Mielke, 1993, pp. 134–158; Firla, 1994, pp. 60–94; Raum, 1997). Except for Raum's article in English, the publications of German researchers are unknown in South African historiography. The French researcher Fauvelle-Aymar (2002, pp. 237–248), who discusses only the abbreviated French translation of Kolb's work and leaves the German edition unconsidered, takes a favourable view of Kolb in his study of the

the Khoikhoi and tries to *persuade* his readers to adopt a more favourable image. He had to attempt to solve an enormous rhetorical problem. He tried to replace a well-established discourse on the Khoikhoi as an abject form of life with a more sympathetic perspective. As he succeeded in this endeavour to a high degree, his book had great significance for Cape ethnography. It would be possible to divide the ethnographic discourse on South Africa up to 1800 into the pre-Kolb and post-Kolb periods.

Peter Kolb at the Cape

Peter Kolb lived in the Cape Colony between 1705 and 1713. He had travelled to the Cape on instruction of the Prussian *Geheimrat* Baron von Krossick to carry out astronomical observations. After the death of the baron in 1707 Kolb received no further financial support. He now had to earn his own livelihood and succeeded in being appointed secretary of the colonies of Stellenbosch and Drakenstein in 1711, for which he moved from Cape Town to Stellenbosch. He stayed in this post until he was discharged because of his blindness in 1712. In 1713 he returned to Germany, where he was appointed head of the gymnasium in Neustadt an der Aisch in Northern Bavaria by the viscount Georg Wilhelm of Brandenburg.[9] Out of gratitude Kolb, as "unterthänigst-gehorsamster Diener" (most humble and obedient servant) then dedicated his book to the viscount, who had organised a post as civil servant for Kolb's father before.[10]

A tradition of rumour-mongering surrounding Kolb's stay at the Cape dates back as far as the eighteenth century. Not only did almost every subsequent author take pleasure in pointing out errors in Kolb's book, but there were attempts to attack the author's reputation in other ways as well. Kolb did not see much during his stay at the Cape, Cape residents told Nicolas Louis de la Caille (1713–1762) and Otto Friedrich Mentzel (1709–1801). Instead of doing astronomical studies, he had been sitting

representation of the Khoisan, even if he readily believes the story that Kolb committed plagiarism (see below).

[9] The biographical information comes from Kolb (1719) and from a short overview by Wolfschmidt (1978). Van Gelder (1997, pp. 266–268) also supplies biographical information about Kolb, including a photograph of the house where he was living when he wrote *Capvt*.

[10] Kolb (1719, "Dedicatio").

smoking and drinking under the trees of the Company's Garden at the Cape. No wonder that his patron stopped his allowance. After his career as an astronomer had been terminated due to dereliction of duty and he had entered the service of the VOC as secretary of the colonies of Stellenbosch and Drakenstein, he was discharged for persistent laziness and blindness, which had been caused by alcohol abuse. Back in Europe, according to La Caille, he published a description of the Cape in Dutch in order to help the Cape colonists in their opposition to the regime of VOC governor Willem Adriaan van der Stel.[11] He did not shrink from ploughing with someone else's heifer either, as a large part of his work was allegedly based on a lost description of the Khoikhoi by the retired Secretary of the Cape Council of Policy, Johan Willem van Grevenbroek (1644–1726).[12] According to the same well-informed circles at the Cape, the kind-hearted Van Grevenbroek had entrusted this document to Kolb, who then unscrupulously plagiarised it after the former's death. Quite in keeping with his nature, Kolb himself was too lazy to go travelling and studying the practices of the Cape natives and vain enough to appropriate someone else's merits.[13]

All sorts of inconsistencies suggest that most of this was malicious gossip.[14] Van Grevenbroek died in 1726, the same year as a Kolb, long after Kolb's German book had appeared in 1719, at which time the troubles around governor Willem Adriaan van der Stel had also been long forgotten. The only irrefutable fact is that Kolb, as he himself also says a few times, may have made use of the notes of his good friend Van Grevenbroek, who also lived in Stellenbosch.[15] And indeed there are a few conspicuous similarities between Kolb's book and an

---

[11] 1664–1733, governor of the Cape Colony from 1699 to 1707.

[12] La Caille (1763, pp. 315–353).

[13] Mentzel (1785, pp. xxx–xxxi) writes: "Die angenehmsten Sommer-Abende schienen ihm in dem groszen Garten der Kompagnie anmuthiger, und er divertirte sich daselbst mit seinen neuerworbenen Freunden öfterer als mit seinem Tubis auf dem Observatorio." (The most pleasant summer evenings seemed to him more attractive in the spacious Company gardens, and he amused himself there with his new-found friends more often than with his telescope in the observatory.)

[14] See Schapera's refutation of the allegations (Schapera & Farrington, 1933, pp. 162–164).

[15] References to Grevenbroek in Kolb (1719, p. 353): "Joh. Wilhelm de Grevenbroek, meinem besonderen Freund" (my special friend) in the part dealing with the origins of the Khoikhoi; "Annotationes" by Van Grevenbroek concerning the worship of the mantis (Kolb, 1719, p. 416). For the enigmatic Van Grevenbroek see Van Stekelenburg (2001).

unpublished letter by Van Grevenbroek about the Khoikhoi. However, there is no proof of large-scale plagiarism.[16]

Kolb's book

Kolb wrote his book in Europe after his return from South Africa. He had to rely substantially on memory because part of his notes had been lost.[17] The book is divided into three parts, each with about the same number of pages. The "Erster Theil" (first part) (pp. 1–346) discusses the "*Physicalia*", the natural environment of the Cape. "Der Zweyte Theil" (pp. 347–582) deals "einig und allein von den Hottentotten" (solely and only with the Hottentots), the original inhabitants of the Cape. In "[d]er) Dritte Theil" (pp. 583–846) he describes "[d]ie Europäischen Colonien an dem Vorgebür[ge] der guten Hoffnung" (the European colonies at the Cape promontory). The parts are subdivided into letters, each dealing with a separate subject. The whole is introduced and concluded with a description of the voyage to the Cape (pp. 1–50) and the return voyage to Europe (pp. 840–846). It is noteworthy that an almost equal number of pages is allocated to the different subjects: if the outward and return voyages (pp. 1–50, pp. 840–846) are ignored, then 237 pages are devoted to the natural environment, 235 to original inhabitants and 256 to colonial society.

Kolb arranged most of his material thematically. Only the third part contains a chronological account of political developments during his stay at the Cape. The rest of the material is arranged by subject, first in larger sections (nature, indigenous population and colonial society) and then in subdivisions by subject. This systematic presentation of information is in agreement with the recommendations of the *ars apodemica*, the art of travel, which recommended that information collected

[16] As the text by Van Grevenbroek (his "Annotationes", as Kolb called them) have not survived, all suspicions of plagiarism are based on guesswork and on Van Grevenbroek's letter about the Khoikhoi (Van Grevenbroek, 1933). Like Kolb, for example, Van Grevenbroek (1933, pp. 208–209) derives monorchy among the Khoikhoi (see below) from Jewish circumcision practices. On the other hand, Van Grevenbroek's letter is much more concise than *Capvt Bonae Spei Hodiernvm* and Van Grevenbroek's scholarly letter lacks all the systematics which are so much in evidence in Kolb's work.

[17] Kolb (1719, p. 829) rarely refers to his own notes—"Annotationes"—which he made during his stay in South Africa and which provided part of the material for his work. However, some of these notes were lost (Kolb, 1719, p. 173).

during a journey should be presented in a systematic arrangement to make it more accessible.[18]

Kolb himself said that his book had been prompted by annoyance. In the first place he was annoyed at earlier publications which gave the Cape only passing attention, the books of passers-by. His model was *An historical relation of Ceylon* by Robert Knox of 1681.[19] Like Knox had written a monograph about Ceylon after many years' residence on the island, Kolb wanted to give a description of the Cape that was as complete as possible. But whereas Knox confined himself to nature and the indigenous population, Kolb also described in the third part of *Capvt* the colonial society of the Cape, and unlike Knox he presented his material as a series of letters addressed to a good friend. The arrangement of the text in letters enabled him to write his account in a style that differed significantly from the usual ornate style and to approximate a style suitable for (neutral) historical description. In this way, style had to be an extra guarantee of the truth of the description, since a letter to a good friend—the genre that Kolb employed in *Capvt*—was, according to him, characterised by frankness and contrary to the artificial style and fabrications of the novel.[20] For this he had different recent precedents, of which he mentioned three by name: the accounts by "P. le Comte", "De la Hontan" and "Wilhelm Bossmann".[21] However, it seems that later publishers did not consider the letter form used in *Capvt Bonae Spei Hodiernvm* to be very successful, because in all later versions and translations it was omitted. Actually, Kolb had already foreseen this possibility. By making small changes—inserting the word "Capitel" (chapter), deleting set phrases used at the start and at the end of letters—an editor could readily modify the stereotypical letters into chapters.[22]

---

[18] See Stagl (1979, pp. 611–638) in this regard.

[19] Kolb (1719, "Vorrede. An den geneigten Leser"—Preface. To the gentle reader). A modern edition of Knox (1681) is Sarapamadu (1956–1957).

[20] *Capvt*, "Vorrede. An den geneigten Leser".

[21] Kolb (1719, "Vorrede. An den geneigten Leser"). He refers to the following authors and their works, in the same order: Le Comte (1696); De Lom D'Arce Lahontan (1703); Bosman (1704). He also mentions a Dutch travel account of a journey to Italy that he read during his stay at the Cape, but of which he had forgotten the title and the author. I have been unable to trace the title of this account.

[22] Kolb (1719, "Vorrede. An den geneigten Leser"). In the Dutch translation of 1727 (see annex 1) the omission of the letter conventions is the only important, textual intervention. As the letter conventions are a peripheral aspect of Kolb's work, the intervention did not affect the contents.

Kolb had more ambitions than only wanting to give a complete description of the Cape. In other authors, he had "found much useless, misleading and totally incorrect things written about the Hottentots and their country", which he wanted to correct in his own book.[23] Earlier authors had allowed themselves to be misled by colonists or unreliable informants during their brief stay. Kolb, however, said that he had proceeded extremely carefully in collecting his information. Everything he had to offer was based on his own observations (autopsy).

Kolb employs a rigid composition in order to make his material accessible to the reader. The work as a whole is framed by accounts of his outward and return voyages. Within this frame, there is a movement from the larger to the smaller. After the geographic position of the Cape has been established, a description of the geography follows that Kolb calls the "äusserliche Schale" (outer shell).[24] Once this external shell has been peeled off, it is time to consume the rest. In accordance with the rules of the *ars apodemica*, the technique of writing travel accounts that recommended an orderly presentation of information, he proceeds systematically, in a methodical manner, since he has to present an enormous amount of information (p. 136). The system he follows is either of his own making or the most self-evident. He adopts the familiar distinction of the three kingdoms of nature, and where necessary he inserts an excursion ("Ausführung") to deal with what falls beyond this classification. He does not use a special classification such as was developed by contemporary natural historians, because he is not writing a "speciale Historie der Tiere"[25] (special history of animals). In this modest structure the alphabet suffices as a mechanism for classifying plants and animals within the traditional arrangement in main groupings (for example: terrestrial animals with feet, birds, marine animals, snakes and insects).[26] A classification of Kolb's own fabrication is that of Cape water in terms of colour, taste, temperature and weight.[27] Naturally, this system increases the accessibility of the material for the reader—who in addition also has an index and a comprehensive table

[23] "[V]iele nichts nützige/ irrige und gantz falsche Sachen/ von den Hottentotten und ihrem Lande / [...] aufgeschrieben gefunden" (Kolb, 1719, "Vorrede").

[24] Kolb (1719, p. 109).

[25] Kolb (1719, p. 137).

[26] Kolb (1719): terrestrial animals with feet (pp. 136–173), birds (pp. 173–190), marine animals (pp. 190–211), snakes and insects (pp. 212–226).

[27] Kolb (1719, pp. 74–75).

of contents at his disposal—and it makes matters easier for the author as well. A separation of subjects in the discourse prevents the author from getting lost in a labyrinth ("Irr-Garten").[28] The order does not give the impression that the division is founded on some sort of hierarchy, though. The system serves especially to prevent information about a given subject being scattered all over the book. The presentation ensures that everything carries equal weight, which is clearly expressed in the equal number of pages allocated to the main subjects—nature, the indigenous population and the colony.

As a result of Kolb's pursuit of completeness, he also tackles subjects of which he knows little. With regard to zoology, for example, Kolb has few pretences, and in consequence he accomplished little in this field, according to the historical zoologist Rookmaaker.[29] On the one hand, the description of nature meets the physico-theological principle of glorifying God's greatness by a praising the diversity of His creation.[30] On the other he wants to prove that as a "curieuser und verständiger Reisender" (inquisitive and intelligent traveller), in accordance with the tradition of the *ars apodemica*, he has paid attention to everything.[31] Thus he was able to prove among other things that the story that the pelican feeds its chicks with blood from its breast is not true.[32] Jonah, he says, was not swallowed by a whale but by a shark. Kolb was able to ascertain personally that nothing more than an average arm could pass through the throat of a whale and that it was therefore too small to allow a man to get in and out again. Moreover, it was known from literature that people had been found in the bellies of sharks. In one shark, a fully armed man had even been found. Unless we assumed that Jonah had landed in the belly of a whale through a miracle of God, he must have been eaten by a shark.[33]

The description of nature in the first part must also serve as an introduction to the description of human society in the next two parts.[34] Here Kolb distinguishes especially between a description of the indigenous inhabitants and of colonial society. As remarked earlier, this

[28] Kolb (1719, p. 348).

[29] Kolb (1719, p. 211); cf. Rookmaker (1989, p. 29). Kolb also says that he has little knowledge of botany and geology (Kolb, 1719, pp. 226, 237).

[30] Kolb (1719, p. 109).

[31] Kolb (1719, p. 136).

[32] Kolb (1719, p. 182).

[33] Kolb (1719, pp. 198–199).

[34] Kolb (1719, p. 347).

distinction seems to me to have been inspired rather by the *dispositio* (rhetorical disposition) of his material—preventing him and the reader from getting lost in a labyrinth—than by an ideological intention to isolate the Khoikhoi.[35] This would not have served Kolb's attempt to rehabilitate the Khoikhoi either, as I will show later.

The second part is ethnographic and describes the appearance, customs and material culture of the Khoikhoi. In as much as this part is historicising, it is based especially on Kolb's hypotheses on the origin of the Khoikhoi. The third part contains a description of the administrative structure of the colony and a brief history of the European presence at the Cape, emphasising the events that Kolb witnessed during his stay there. The main subject of this part is a detailed report on the discord between the Cape freeburghers and governor Willem Adriaan van der Stel in 1706, in which Kolb takes sides against the governor.

"Einig und allein von den Hottentotten"

Despite the 'horizontal' classification in which nature, indigenous population and colony are given equal space, the second part—which deals "only and exclusively with the Hottentots"—is nevertheless the most interesting part of the whole book. The central position occupied by the Khoikhoi is revealed in several ways: they are depicted on the frontispiece of *Capvt Bona Spei Hodiernum* (fig. 8), take pride of place in the title, and moreover Kolb mentions in the "Vorrede" (preface) the desire to correct the false statements about the Hottentots made by other authors as a reason for writing *Capvt Bona Spei Hodiernum.* His contemporaries also showed interest especially in this part. Reprints of the parts about the indigenous population appeared in English and French, but not of the part about nature at the Cape (see annex 1). The third part, containing a description of the colony, remained largely untranslated into English. When eighteenth-century *philosophes* referred to Kolb, they used exclusively Kolb's information about the Khoikhoi.

[35] In some places he indicates that a certain subject had to wait for the sake of the systematics (Kolb, 1719, pp. 348, 584, 730). I mention this because from a postcolonial perspective a tendency exists to promptly interpret separations of this kind as attempts to 'other'.

Kolb's intention with this central part of his book, as he states in his "Vorrede", is to take the Hottentots' side against false accusations.[36] This put him in the position of the speaker who had to persuade his public, because this public had the extremely negative picture that was painted in the existing literature such as the assessment of the Hottentots by Wouter Schouten that was quoted at the start of this chapter. Kolb's book was to become a rhetorical as much as a scholarly tour de force.

We can therefore formulate Kolb's strategy in rhetorical terms.[37] What Kolb envisages in the second part of his book is to replace the *vituperatio* (deprecating description) prevailing in the existing literature on the Cape with a *laus* (laudatory description) without creating the impression that he was acting as an "advocate".[38] He presents his description as a factual narrative, as (objective, factual) historiography ("die Pflicht eines Historien-Schreibers"—the duty of a historiographer),[39] which obliges him not to conceal completely acts of the Hottentots which were indisputably sins in the eyes of Europeans, such as the killing of twins (fig. 9).[40] After all, this would detract from the credibility of his narrative and would not befit his role as historiographer. In his account Kolb attempts to replace the *vituperatio* with a laudatory description of a subject that has its favourable as well as its bad characteristics, but of which the good characteristics are emphasised as much as possible in order to present the most favourable picture (*laus* of an *amphidoxa* or *dubium*).[41] The rhetorical quality of his description also permeates his way of reasoning. The argument is in the form of rhetorical induction,

[36] "Hernach hat mich dazu veranlasset und bewogen / weil bey einigen Scribenten / viele nichts nüszige / irrige und gansz falsche Sachen / von den Khoikhoi und ihrem Lande / habe aufgeschrieben gefunden" (Thereafter gave me reason and motivated me because in some authors I found much useless, misleading and totally wrong things written about the Hottentots and their country (Kolb, 1719, "Vorrede").

[37] Kolb was trained in rhetoric (Wolfschmidt, 1978, p. 8), like everybody of his generation with a university background. There is extensive literature on the influence of rhetoric on writing in Germany (and Europe) up to the end of the eighteenth century (see among others Dyck, 1969; Witstein, 1969, pp. 3–134; Lausberg, 1960).

[38] Kolb (1719, p. 440).

[39] Kolb (1719, p. 440). Historiography was a broad concept in the eighteenth century that could be used for any kind of factual description. Voltaire described historiography ("histoire") in the *Encyclopédie* as "an art that transmits the facts in writing" and "an account of facts presented as true" (Duchet, 1985, p. 8). See also: Seifert (1976).

[40] Kolb (1719, pp. 440–446).

[41] Lausberg (1960, pp. 57–58, 129–136).

where the conclusion is rendered credible by examples.[42] A crucial element of Kolb's defence is a refutation or mitigation of the accusations (*refutatio*) brought against the Hottentots.[43] He does this in two phases.

First of all, in his Preface he casts doubt on the expertise of all those who wrote about the Cape before him. Referring to the reverend Ziegenbalg,[44] he represents these authors as writers who had been taken for a ride by believing all sorts of nonsense that was told to them about the Hottentots at the Cape. He says that this habit of relying on tales told about the Hottentots by the Cape colonists has led to a distorted picture. Just imagine, says Kolb, that a Hottentot were able to read in Germany what was being written about his people; he would be just as annoyed as the abbot Gregorius of Abyssinia when he read the rubbish that was being written about his country in Germany. Everything that Kolb writes himself, on the other hand, is based on *autopsy* (Kolb uses the Greek alphabet for this word) "or [my] own observation, and personal attendance".[45] This had taken him "many exhausting journeys, thousands of drops of perspiration and indescribable difficulties"—this is something that is not all that evident in his text, as he arranged his book systematically and omitted inessentials such as the circumstances in which he gathered his knowledge from the text as far as possible.[46]

Kolb presents himself as reliable, as someone who has made all efforts to collect reliable empirical information. He writes in a simple style, he says, as in letters one writes to a friend, not in the inventive

---

[42] Milton (1989, pp. 52–53).

[43] Lausberg (1960, p. 540) points out the anti-traditional, 'enlightening' character of the *refutatio*: "Die *refutatio* [...] ist der Versuch der Widerlegung eines von der Tradition überlieferten und von der Traditionspartei anerkannten historischen oder mythologischen Faktums. [...] Die *refutatio* ist eine Waffe der historischen Kritik und eine Waffe der Aufklärung im Weltanschauungskampf." (The *refutatio* [...] is the attempt to refute a historical or mythological fact passed on by tradition and recognised by the traditionalist party. [...] The *refutatio* is a weapon of historical criticism and a weapon of enlightenment in the battle between world views.)

[44] Ziegenbalg (1715).

[45] "[...] oder den eigenen Anschauen / und persönlichen Beywohnen" (Kolb, 1719, "Vorrede"). This does not mean that Kolb learned this from the new, experimental natural science of Newton and Boyle. The use of the Greek word autopsy (in his Preface, Kolb even uses the Greek alphabet) is more indicative of an older tradition in the natural and cultural sciences that goes back as far as Herodotus. In this tradition special merit was also attached to own observation (Schepens, 1980; Hartog, 1980, pp. 272–275). Besides, Kolb's background in natural science very likely contributed to an empiricist attitude towards the exotic world.

[46] "[...] manchen mühesamen Weg / viele tausend Tropffen Schweisses / und unbeschreibliche Beschwerlichkeiten gekostet". He refers occasionally to the expeditions he undertook in order to gather information (Kolb, 1719, "Vorrede", pp. 523, 576–582).

fashion of the (untruthful) novel.[47] The reader is even asked to put up with Dutch grammatical constructions—a result of Kolb's stay at the Cape, where he learned Dutch in order to be able to converse with to everybody.[48] This statement also serves as a guarantee that, unlike his non-Dutch predecessors, he was able to understand Cape informants well. The Dutch grammatical constructions in his German text—which, incidentally, I did not notice—are also supposed to furnish substantive proof of his long residence at the Cape.[49]

Kolb's second step is a systematic rebuttal or mitigation of all the clichés which were current in the discourse on the Hottentots. Here he always first quotes what some important authors remarked on a certain subject and then gives his own opinions and observations, where the reader (on the basis of Kolb's remarks in his Preface) is obviously expected to attach more credence to what Kolb has to say than to what the quoted authors assert. The result is a recalibration of set elements in the existing discourse on the Hottentots. A good number of opinions are rejected, a few are upheld and almost all of them are re-contextualised, which, eventually, results in a more favourable perspective on the Hottentots. His trump card in this issue is certainly a comparison between the pagan (and therefore morally naïve) Hottentots and Christians. In many respects the Hottentots are found to have better qualities that the Christians, and in the cases where they had bad characteristics these had often been acquired from the Christians. At least these bad characteristics could be found to a similar degree amongst Christians or other peoples that were regarded as being civilised, such as the Chinese or Japanese. We will return to this later.

A continuous thread in the construction of a more favourable perspective is the genealogy of the Hottentots. Following the dialectic pattern, according to which he initially debates the existing literature on a topic, Kolb introduces his thesis on the origin of the Hottentots by first refuting an absurd story. Contrary to the assertion of "Jacob Sadeur" in "*Nouveau Voyage de la Terre Australe*",[50] Kolb maintains that the

---

[47] Kolb (1719, "Vorrede").

[48] Kolb (1719, "Vorrede").

[49] Kolb does regularly take over characteristic expressions in (creolised) Dutch.

[50] De Foigny (1692). This is an imaginary traveller's tale of which Jacques Sadeur is the fictitious author. The passage in question about the indigenous population of the Cape reads as follows: "[...] nos recherches ne servirent qu'à trouver l'origine d'une Nation voisine, que les Européens appellent Caffres, & les Naturels du Pays Tordt; Nous apprîmes donc qu'un homme du Pays aiant élevé une petite Tigresse, devint si familier avec cette bête qu'il l'aima charnellement, & commit un crime infame avec

Hottentots are not "Kaffirs", nor have they sprung from the mating of a man and a tigress.[51] But then what is their origin? Noteworthy, says Kolb, are the similarities between the Hottentots and the Jews and the Troglodytes from North Africa: "Because they have much, when all circumstances are taken into account, in common with the Jews and also a great deal with the Troglodytes. These Troglodytes are ancient African peoples who are the descendants of the Carthaginians."[52] Kolb points out no less than ten similarities with the Jews and seven with the Troglodytes.[53] The table below summarises these similarities (table 1).

On the basis of these similarities he assumes that they are descended from the Jews as well as from the Troglodytes. Their better, 'inner' customs derive from those of the Jews, the 'outer' customs of a lower calibre (such as leaving frail elderly people to fend for themselves) they would have inherited from the Troglodytes, who would be related to the Carthaginians.[54] Ultimately, the culture of the Hottentots is a mixture of all these cultures. As they wandered about, they forgot their origins and their customs and became creolised, as one would say today; but in Kolb's view the constituent elements are still recognisable.

elle, dont il vint un animal demi-homme & demi bête, Monstre qui a donné l'origine à ces Sauvages [the Khoikhoi] qu'on ne peut humaniser. Une preuve tres vraisemblable de cette Histoire, c'est que leur tête & leurs pieds ont de grands rapports avec ceux des Tigres, & leurs corps même sont en quelsques endroits marquetez de tâches, pareilles à celles de ces animaux" (De Foigny, 1692, pp. 53–54). ([...] our research serves only to find the origin of a neighbouring nation, which the Europeans call Kaffirs, and the natives of the Land of Tordt. We learnt that a man of the Country, having raised a small tigress, became so fond of this beast that he loved her carnally and committed a beastly crime with her of which issued an animal that was half man and half beast, a monster from which are descended these savages [the Khoikoi], which one cannot humanise. A very probable proof of this story is that their heads and their feet are very similar to those of tigers and their bodies in some places bear spots which are the same as those of these animals.)

[51] Kolb (1719, p. 351–352).

[52] "Denn sie haben vieles, nach allen Umständen, mit den Jüden und wiederum sehr vieles mit denen alten *Troglodyten* gemein; welche uralte *Africani*sche und von den *Carthaginen*sern herstammende Völcker gewesen" (p. 352).

[53] The similarities with the Jews are taken from various parts of the Old Testament. The main sources of the information about the troglodytes are Diodorus Siculus (90 B.C.–30 A.D.) (Oldfather, 1935, pp. 172–173) and Joannes Bohemus (or Boemus) (Grimm & Wirsung, 1520, pp. xiv–xv, with thanks to Geoffrey Etough for transcribing and sending the relevant passage). The information of Boemus about the Troglodytes corresponds largely to Diodorus. The information of Diodorus Siculus was based on Strabo (16.4.17).

[54] Kolb bases this contention on Hiob Ludolf, *Historia Aethiopica* (1681). I did not have this work at my disposal.

	Similarities with the Jews		*Similarities with the Troglodytes*
1	Many sacrifices	1	They call their children after their pets
2	New moon celebrations	2	They leave frail old people behind to die alone
3	At certain times they may not touch their women	3	Women can intervene when two men are fighting
4	When caught out doing the above, they must sacrifice	4	They can also run fast
5	They eat unleavened ('ungesänert') bread and unsalted food, a reason why they stay away from Christians	5	They have curious funeral rites
6	They practise circumcision	6	They hunt frequently
7	They may not eat asphyxiated animals	7	They live in a hot country
8	They may not eat fish without scales		
9	Their women are not allowed at political meetings		
10	They can give their women letters of divorce		

Table 1. Similarities of the Hottentots with the Jews and Troglodytes

And who could quickly and concisely explain all these similarities? Suffice to say that they correspond to both the Jews and the Troglodytes in many respects. In the circumstances, it is difficult to pick one or the other alternative, as the agreement with the Jews relates to the inner, whereas correspondence with the Troglodytes usually concerns the outer and more gruesome aspects. In my view one can deduce from this that they are ancient African peoples who, because they were always chased away and driven further and further from their original place of residence, are a mixture of the Jews who moved here and of other African, especially Carthaginian, peoples. In the end they wandered to the extreme tip of this country and settled down there and formed a community. Because they had been so many nations, they borrowed [customs] from each other, which resulted in the loss of their own customs. As a consequence of which, one encounters this chaotic condition in which they live.[55]

[55] "Und wer wolte alle diese übereinkommende Stücke so schnell und kurtz beybringen? Es ist genug daß daraus ersehen wird, wie sie so wohl mit den Jüden, als *Troglodyten* in ziemlich viel Stücken eine Ubereinkunfft haben. Es ist unterdessen eine gar schwehre Sache, daß man itzt einer dieser Meynungen beypflichtet; zumal da die Ubereinstimmung met den Jüden zugleich auf das innerliche siehet: da im Gegentheil der *Troglodyten Convenientz*, meist aus äußerliche und dabey offt grausame Sachen

In Kolb's presentation of the origins of the Hottentots their descent from the Troglodytes is always of secondary importance. The more significant similarities, especially the (semi-)religious ones, are those with the Jews. They are essentially Jews who, forced by circumstances, intermarried with other peoples. Because they say among others that their first ancestors were Nôh (the man) and Hingh-nôh (the woman), and does the man's name not resemble that of "Noah"?[56] What was initially only a "credible" genealogy[57] even becomes, a few pages further, a certainty that needs only to be illustrated in the subsequent sections of *Capvt* on the Hottentots:

> But why do I now take trouble to make a matter credible to the reader that is more than certain on its own? Everything will become clear when in the letters following below the circumstances are revealed that will abundantly illustrate and fully explain the whole issue.[58]

From the moment Kolb presents his picture of the genealogy of the Hottentots as a certainty, the rest assumes the form of an argument based on examples. It also appears as if the quantity of these examples had to compensate for the lack of quality thereof. Incidentally, Kolb's statement that the Hottentots were descended from the Jews was not so unusual in the eighteenth century. Since the sixteenth century a theory had existed that the American Indians were actually Jews who had emigrated, and in humanist circles there was also some interest in the origin of the Jewish communities in Barbary, in North Africa.[59]

---

gerichtet stehet. Meines Orts lasse mich bedüncken, daß unwiedersprichlich hieraus zu ersehen, wie dieses uralte *Africani*sche Völcker seyn; die, weil sie immer vertrieben, und je länger je weiter von ihrem alten Sitz verjaget worden, so wohl aus denen dahin gebrachten Jüden, als auch andern *Africani*schen und absonderlich *Carthaginensi*schen Völckern, sich zusammen geschlagen, und endlich an diese äusserste Spitze des Landes begeben, daselbst gesetzet und häußlich in eine Versammlung begeben haben: und weil so vielerley *Nationen* gewesen, so habe immer eine von der andern was genommen, jede aber ihre absonderliche Gebräuche vergessen; daß also nunmehro ein verwirrter Zustand bey ihnen anzutreffen" (Kolb, 1719, p. 353).

56 Kolb (1719, pp. 353–354).

57 Kolb (1719, p. 353).

58 "Jedoch was bemühe ich mich viel, Ihm [the reader of Kolb's fictitious letter] eine Sache glaublich vorzustellen, die an sich selbst mehr als zu gewiß ist? Es wird sich dieses alles klärlich darthun und zeigen, wenn unten und in den folgenden Briefen solche Umstände werden vorgestellet werden, die die gantze Sache überflüssig *illustriren* und hertzlich erläutern" (Kolb, 1719, p. 355).

59 Hallowell (1960, p. 5); Eilberg-Schwartz (1990, pp. 32–36); Horowitz (1992, pp. 309–326).

"Exsecutio Testiculi"

I want to examine a concrete example of such a similarity between Khoikhoi on the one hand and Jews and Troglodytes on the other more closely. It concerns the delicate matter of partial castration ("exsecutio testiculi") among the Khoikhoi. This is a classic topic from the seventeenth-century discourse on the Cape that also makes it possible to compare Kolb's approach with that of his predecessors, the passing travellers, who in the accounts of their travels always inserted an obligatory piece on the indigenous population.

Semi-castration—monorchy—attracted the attention of passing travellers from the start of the seventeenth century. The first to remark on it was probably the Englishman Edward Middleton, who states in 1607 that some Khoi men "have but one stone".[60] Once the phenomenon had been observed, subsequent authors tried to collect more information. Attempts were made to determine whether the left or the right testicle was excised, at what age this was done and to what end. All differences aside (there is not even agreement on whether the left or right testicle was removed), what all the different authors have in common is that the subject is always dealt with cursorily.[61] Kolb is the first to investigate this remarkable practice thoroughly and to compare the prevailing opinions with his own findings. He dedicates almost the entire sixth letter in the second part to it: "von den Opfern der Hottentotten; oder wie sie es zu nennen pflegen / vom Anders-machen: absonderlich aber von der *Exsecutione Testiculo* &" (Of the sacrifices of the Hottentots, or as they call it / of making different: particularly of the exsecutione testiculo &).[62]

As regards the existing literature he confines himself to Saar, Tachard, Böving and Vogel,[63] who are quoted for their opinions on the age at which the operation was done (the general opinion was eight or nine years) and the reason why it was done. Kolb rejects these opinions in favour of his own findings, which are presented in such a way that the reader must clearly notice that they are based on autopsy.

---

[60] Raven-Hart (1967, p. 33).

[61] See Raven-Hart's index (1967; 1971).

[62] Kolb (1719, pp. 420–433).

[63] Johann Jacob Saar (1662); Guy Tachard (1686); Johann Georg Böving (1712); Johan Wilhelm Vogel (1690).

The point of departure of his description is that he has indeed established—he has "visitiret" (examined) no less than a hundred men—that all men have no more than one testicle.[64] He did not shrink from bribery to obtain his information:

> I have [...] examined many people out of scientific curiosity and persuaded them with a piece of tobacco or a penny to take off their kul-kros, i.e. the piece of leather that they wear in front of their *membro virili*, thus giving me full freedom to see everything I wanted to see and wanted to know of them.[65]

Kolb is the only author about the Cape who also describes the operation whereby, according to him, the left testicle is removed.[66] And before he does this, he excuses himself to the reader with the Latin proverb "*naturalia nun sunt turpia*" (natural things are not abhorrent). He is not going to follow the example of those "Wohl-Rednern" (flatterers) who cloak such subjects in obscure language; this matter is too new and too important for this. Kolb is going to give a clear and accurate description of what he has seen.

For the operation the patient is laid on the ground and tied down. Others put their knees on his arms, feet and chest. Then an old Khoi approaches with a sharpened bread-knife—the "Hottentotische Kappaunenmacher" (Hottentot capon-maker), as Kolb calls him (fig. 10). According to Kolb, this surgeon is accorded the same respect as a rabbi among the Jews.[67] Then the operation is described:

> He takes the testicle in his hand, makes an incision about as long as one and a half knuckle-bone and squeezes out the testicle, which he then cuts through and off on the back side, not at the veins or the bladder or other vessels but immediately at the end.[68]

---

[64] Kolb (1719, p. 420).

[65] "Ich habe [...] sehr viele aus *Curiosité* besichtiget, und sie gegen ein Stück Toback oder ein *Dubbeltje* dahin gebracht, daß sie ihre *Kul Kross*, oder das Läpplein Fell, welches sie über dem *Membro virili* tragen, hinweg gethan, und mir dadurch volle Freyheit gaben, nach allen zu sehen, was ich nur wolte, und von ihnen zu wissen begehrte" (Kolb, 1719, p. 423).

[66] Kolb (1719, p. 423).

[67] Kolb (1719, p. 422).

[68] "[...] fässet er den *Testiculum* in die Hand, machet eine Oeffnung, ungefehr anderthalben Glieds-lang und drücket den *Testiculum* heraus; welchen er nachgehends hinten, nicht an den Geäder oder Harn—auch andern Gefässen, sondern gleich zu Ende desselben durch und abschneidet" (Kolb, 1719, p. 422).

While the operation is in progress, a sheep is slaughtered. The kidney fat of the sheep is mixed with *buchu* and other herbs and rolled into a ball about as large as the testicle was ("ohngefähr als der Testiculus gewesen") and is put in the place of the excised testicle, after which the old man sews up the wound with a sharpened bird bone and the nerve of an ox or a sheep. When the operation has been completed, the old man urinates on the patient, which is also shown on the accompanying illustration. After two or three days of suffering, the patient happily runs about again.[69]

In addition to describing the operation, Kolb also takes up the question at which age the operation is done and why it is done. He has had to correct his own opinion as far as the age is concerned, he says. Initially he wrote to a friend that it was done at the age of eight or nine, as was asserted by most of the other authors writing about the Cape, but he also knows of an operation on an eighteen- or nineteen-year-old boy, a certain Jan or Johannes.[70]

Finding an explanation for the partial castration is more difficult. In the older literature the general opinion, which is also shared by the four authors quoted by Kolb, was that it was done in order to be able to run faster.[71] Some authors also thought that it was intended for birth control or to inhibit the sexual urge.[72] Kolb denies that the purpose of the operation was to increase athletic prowess. In his view one had to look for the cause in marital practices: "weil sie vermög ihrer Gesetze, nicht bey einer Hottentottin schlaffen können, oder mögen, so lang sie 2. *Testiculos* haben" (because according to their laws they cannot or may not sleep with a woman as long as they have two testicles).[73] A woman who slept with a man still in possession of a full complement of testicles

[69] Kolb (1719, p. 423).

[70] Kolb (1719, p. 421).

[71] Besides Saar, Vogel, Tachart and Böving quoted by Kolb, many other authors also thought that the partial castration helped one to run faster, a characteristic of the Khoikhoi that had always enjoyed particular attention. Here I made use of the anthologies of early modern descriptions of the Cape by Raven-Hart (1967, 1971): De Beaulieu (Raven-Hart, 1967, p. 101), Travernier (Raven-Hart, 1971, p. 179), Nieuhoff in imitation of Travernier (Raven-Hart, 1967, p. 19), Schreyer (Raven-Hart, 1971, p. 125), Leguat (Raven-Hart, 1971, p. 433), Bogaert (Raven-Hart, 1971, p. 484).

[72] Birth control: Thomas Herbert (Raven-Hart, 1967, p.123), Schreyer (Raven-Hart, 1971, p. 124), Ovington (Raven-Hart, 1971, p. 396), Langhansz (Raven-Hart, 1971, p. 406). Sexual inhibition: *Klare Besgryving van Cabo de Bona Esperanca* (Raven-Hart, 1971, p. 52), Ovington (Raven-Hart, 1971, p. 396).

[73] Kolb (1719, p. 424). In the literature on the Cape the connection with marital practices was made only by Bolling (Raven-Hart, 19 of 71, p. 147).

was regarded as a whore. On the other hand the operation was also necessary for admittance to the community of men. According to Kolb, it is—in modern anthropological terms—a rite of passage. The man who had not had the operation did not become a man but remained a "Kutsire", an epithet meaning something like a milksop.[74]

Kolb's description serves two purposes. On the one hand it is a statement of an ethnographic fact. His eyewitness account has to disclose the facts of the matter as regards what had become the infamous question of partial castration among the Hottentots. To this end he confronts the existing discourse with his own findings. Because of the accuracy of the description and because the reader is repeatedly reminded that the information was obtained by personal visual observation, Kolb's representation has to be more credible than what little can be found in the existing literature. He himself states in his Preface that the reader will not find such comprehensive information as contained in his book anywhere else. The intended result is that all more or less authoritative descriptions of the partial castration up to that point have to be rejected in favour of Kolb's description.

The Jewish descent of the Hottentots

The account also plays a part in his theory about the origin of the Hottentots. According to Kolb, this is a religious ritual, and therefore this is further proof that the Hottentots have a religion because a sheep is slaughtered.[75] As the ritual is comparable to that of circumcision among the Jews,[76] it is also proof of the origin of the Hottentots. According to Kolb, it is not "unglaublich" (unbelievable) that the Hottentots have forgotten the original Jewish way of circumcising. They could also have inherited this custom from the Egyptians and the Troglodytes, as can be found in Johannes Bohemus (1485–1535) and Diodorus Siculus (approx. 90–30 BC.)[77] However, Kolb creates the impression that it was precisely the influence of the Troglodytes that caused the original Jewish circumcision to be degraded to semi-castration.[78] Even

[74] Kolb (1719, p. 426).
[75] Kolb (1719, p. 426).
[76] Kolb (1719, p. 426).
[77] Oldfather (1935, pp. 172–172).
[78] "Es ist dahero nicht wohl anders zu muthmassen, es müssen auch viele *Troglodyten* unter den *Hottentotten* anfänglich gewesen seyn, welche diese Weise der Beschneidung

when the details are contradictory, he seems inclined to adhere to the original thesis that the customs are of a purely Jewish origin and were altered by Troglodytic influence. The essence is Jewish, the obscuring *accidentalia* are Troglodytic (or perhaps even Egyptian).

One may ask why Kolb makes such an effort to explain the customs of the Hottentots as being primarily of Jewish origin. If needed, he even violates the rules of logic. According to the sociologist of religion Chidester he does this to put them in a poor light. Chidester argues as follows. While Hottentots, according to Kolb, do have a religion, which makes them more human in the eyes of Europeans, this religion is—like Judaism—especially an opposite of Christianity. Like the Jews, they were also unwilling to give it up.[79] This argument seems very improbable.

As I said earlier, Kolb actually wants to protect the Khoikhoi/Hottentots against their European critics. He says this in so many words in his Preface and thereafter proceeds to put most of their customs in a favourable light. This is, for example, evident in the letter in which he presents the moral balance of their virtues and vices, "Von der Hottentotten allgemeinen Tugenden und Lastern: absonderlich von ihrer Liebe zur Gerechtigkeit" (On the general virtues and vices of the Hottentots: especially their love of justice).[80] In figures, he can list as many virtues as vices (nine per category), but the difference is that most of the virtues are presented as unique characteristics of the Hottentots, the presence of which among these heathen should shame the Christians (who should know better), while the vices are extenuated by saying that one will find most of them among Christians as well.[81] Even the habit of killing the girl in the case of newborn twins, if one or two of them are girls—something that seems to count heavily against the Hottentots—is still to some degree mitigated by saying that one also finds this horrible practice among civilised peoples such as the Japanese and the Chinese.[82] And in any case, the Hottentots would never sell their dead children to Europeans.[83] Even in this case, which

oder Ausschneidung unter ihnen fort gepflanßet haben" (It can therefore hardly be suspected otherwise than that there must initially have been many Troglodytes among the Hottentots who propagated this form of circumcision or excision among them) (Kolb, 1719, p. 425).

[79] Chidester (1996, p. 54).

[80] Kolb (1719, p. 540–555).

[81] Kolb (1719, p. 546).

[82] Kolb (1719, pp. 445–446).

[83] Kolb (1719, p. 446) states that he tried to buy a dead Khoikhoi baby for a collector in Amsterdam.

is extremely unpalatable in the European view, he attempts to present its reprehensibility in the most favourable light. As a "historiographer" one could not keep quiet about these wrongs, but one could make them less odious by producing comparable examples from the civilised world. In my view, the thesis about the lineage of the Hottentots serves the same purpose: redirecting the *vituperatio* of the prevailing discourse into praise. In doing so, reprehensible facts must be acknowledged, but at the same time portrayed as favourably as possible (*laus* of a *dubium/amfidoxa*—praise of a topic which has many good qualities and a few less agreeable ones).

As pointed out earlier, a substantial number of habits of the Hottentots correspond to those of Jews and Troglodytes, which according to Kolb justifies the hypothesis that they are descended from these peoples. The higher, 'inner' customs are creolised Jewish customs and the less pleasant 'outer' customs Troglodytic. The latter does not help very much to raise the status of the Hottentots, even if he does accept on the authority of Hiob Ludolf (1624–1704) that the Troglodytes in their turn were descended from the Carthaginians, which by a detour via the Troglodytic line still provides them with a civilised Mediterranean ancestry. The Jewish descent of the Hottentots is more important. Kolb is inclined to put this forward as the core of the customs of the Hottentots. Miscegenation has adulterated the Jewish nucleus. This also presented the best possibilities for rehabilitating the Hottentots. From the sixteenth to the nineteenth century the image of the Jews in Protestant Europe was generally quite favourable, even to the point that many historians of religion speak of philosemitism.[84] An association of the Hotttentots with the Jews therefore had to improve their image quite substantially. Through the link with the Jews (and the Troglodytes) they also lost their isolated position and became connected to the history of humankind. At the same time they retained their undeniably exotic image, so that 'the truth' was not violated.

That this was indeed interpreted in this way in the eighteenth century appears from a remark by one of the Kolb's fiercest critics,

[84] See: Katz (1992, p. 334); Gritsch (1994, pp. 202). In 1700–1701 Kolb briefly studied oriental languages (read: Hebrew) under the leading Hebraist August Hermann Francke in Halle (Wolfschmidt, 1978, p. 9). In the eighteenth century Halle grew into the centre of missionary work among the Jews, which at the least points to a special interest in Judaism (Vogt, 1998, pp. 21–22). It is quite remarkable that the historian of religion Chidester does not anywhere refer to the firmly established tradition of philosemitism in North-Western Europe to which Kolb can be linked.

Otto Friedrich Mentzel.[85] Mentzel finds Kolb's comparison between Hottentots and Jews ridiculous. One can, says Mentzel, just as well point out similarities between Germans and Jews, but this does not mean that the Germans are descended from the Jews. Furthermore it is improbable that the Jews would forget the Torah, and as far as cleanliness is concerned it is difficult to imagine a greater difference than between the cleanliness of the Old Testament Jews and the filthiness of the Hottentots, who—like dung beetles in excrement—constantly live in filth.[86] Everything has been dragged in by the hair, including the comparison with the Troglodytes.[87] The Hottentots are not related to anyone, Mentzel says. They split from the rest of humankind immediately after the Babylonian confusion of tongues, or—and then Mentzel comes up with an alternative lineage—they are the descendants of children who survived a shipwreck, which explains why they are so childlike in all respects.[88]

Conclusion

Kolb's hypothesis about the origin of the Khoikhoi did not have any followers, although the attention of the reader of the Dutch translation is specifically drawn to it.[89] His project as a whole, however, was successful. Kolb succeeded in specialising the discourse on South Africa. He was the first to publish a monograph on the Cape. Thus he elevated the Cape to a theme to which a hefty book (or two) could be dedicated. Moreover, he introduced for the Cape a method of critical research in which the researcher compared the literature with his own findings instead of blindly repeating clichés, as used to be the custom. After the publication of Kolb's *Capvt* the older literature on the Cape lost most of its value, and on Kolb's authority it was also possible to give a favourable representation of the Khoikhoi. From that period on, reference was usually made to Kolb. Kolb's catalogue of the manners

---

[85] Mentzel (1785–1787), pp. 449–463.

[86] "[...] als zwischen der Reinlichkeit der Juden des alten Testaments und der Unfläthigkeit der Hottentotten, welche wie die Roßkäfer in dem Mist, in einem bestëndigen Unflathe leben" (Mentzel, p. 454).

[87] Mentzel (1785–1787, p. 454).

[88] Mentzel (1785–1787, pp. 459–463).

[89] *Naaukeurige en uitvoerige beschryving van Kaap de Goede Hoop* (Kolb, 1727).

and customs of the Khoikhoi became the programme for subsequent research.

Kolb's natural sciences background may well have played a part in his critical attitude towards the established image of the Khoikhoi. As an astronomer he was used to collecting empirical data.[90] Moreover, astronomy was a branch of science that was undergoing revolutionary changes in Kolb's Times.[91] Jupiter's moons, for example, which Kolb's Prussian patron had requested him to study at the Cape, had been discovered by Galilei not so long before, and it took Galilei a considerable effort to convince the scientific world of their existence. Kolb had grown up with scientific paradigm shifts, and perhaps it was for this reason that he had very little respect for the empirically poorly substantiated discourse on the Cape, as he indicates in his Preface. This empirical and critical scientific attitude enabled him to transform the discourse on the Cape and the Khoikhoi, thereby setting a standard for following generations.

With his emphasis on factual, observation-based evidence Kolb was 'modern' in his own times. But he also had a traditional humanist side that comes to the fore in his frequent quotations of classical, humanist and Christian sources to add weight to his arguments. This is also apparent in his argument that the Khoikhoi are descended from Jews and Troglodytes. Arguments that had to prove descent from the Jews were taken mainly from the Old Testament; descent from the Troglodytes could be substantiated by referring to Diodorus Siculus and the sixteenth-century armchair scholar Johannes Boemus or Bohemus. Even the authority of Herodotus (485–425 B.C.) and Pliny (23–79 A.D.) is occasionally invoked. Kolb therefore found truth not only through own observations, but also derived it from authoritative sources.[92] The new empirical science, however, advocated the primacy of sensory observation. To mark this, the Royal Society of London adopted in 1662 the slogan "Nullius in verba" (on the words of no-one, or: take non-one's

[90] Three manuscripts with Kolb's astronomic observations at the Cape have survived (Wolfschmidt, 1978, p. 23).

[91] See Kuhn (1970) in this regard.

[92] For the use of antique authorities in early modern science see Dear (2001, pp. 30–48). The new world was initially also brought into line with the Bible and authoritative sources from Antiquity (see Grafton, 1992). At the end of the eighteenth century antique sources such as Ptolomy and Leo Africanus were still invoked in the geography Western Africa by the leading English cartographer James Rennell (1742–1830; cf. Bassett & Porter, 1991, pp. 377–378).

word for it), which means among other things that sensory observation takes precedence over book knowledge.[93]

Kolb became the standard not only for researchers who occupied themselves specifically with South Africa. In eighteenth-century speculations on primitive people the Khoikhoi were also sometimes involved, and there too Kolb's representation, directly or through translations and summaries, was an important source of information. After Kolb two types of Hottentots existed in European writings. The monstrous variety harked back to the seventeenth-century descriptions; the 'enlightened' variety was taken from Kolb. These representations are also frequently mixed, as in Voltaire's *Essay sur les moeurs* (1761) and in an article by De Jaucourt on the Hottentots in the *Encyclopédie* (1765).[94] Kolb's enlightened Hottentot gained its greatest prominence in Rousseau's *Discours sur l'origine et les fondaments de l'inégalité parmi les hommes*, (also known as the *Deuxième Discours*) (1775), in which Rousseau (1712–1778) attempted to reconstruct the developmental history of humankind, partly on the basis of a French summary of Kolb's work.[95] In this book Rousseau distinguishes three stages of human development: natural man living in solitude, man as part of a primitive hunting society and finally egoistic, civilised man. According to Rousseau man is happiest in the middle stage of primitive society—the "juste milieu"—and it would therefore have been best had he never left that stage.[96] Rousseau supports his assertions about the first two stages of development with borrowings from descriptions of primitive peoples, including the Hottentots, whom he regards as contemporaneous ancestors of civilised Europeans. Kolb's work, in the form of a summary in the *Histoire générale des voyages* (1748), was his main source of information about the Hottentots.[97] An anecdote

[93] Carey (1997, p. 283).

[94] Fauvelle-Aymar (2002, pp. 247–303). In his *Vorlesungen* Kant used the abbreviated, second German edition of Kolb's work (Firla, 1994). According to Merians (2001, pp. 159–167, 195–196, 200–203) the English translation of Kolb's work was used as a source for satires as well as scholarly works in England.

[95] Compare Rousseau (1964, III, pp. 111–240, 1315) with remarks on the use of Kolb's work by Rousseau. According to notes he left behind, Rousseau took thorough note of a summary of Kolb's work in the *Histoire général des voyages*.

[96] Rousseau (1964, III, p. 171).

[97] Rousseau refers to a summary of Kolb's work twice in the *Histoire générale des voyages*. The first reference is note iv of the *Discours*, in which Rousseau quotes Kolb at length in support of his assertion that primitive people possessed a superior physique. The second instance is the anecdote about the Hottentot who returns to nature and lays down his clothes before governor Van der Stel (Rousseau, 1964, III, p. 200, note xvi). This anecdote supports Rousseau's assertion that the intermediate stage is best

that Kolb had taken over from Guy Tachard (1651–1712)—the assimilated Khoi who turns his back on European civilisation in the person of commander Simon van der Stel and lays his European clothes at the commander's feet[98]—was even depicted on the frontispiece of the first, Amsterdam edition of Rousseau's *Discours sur l'origine et les fondaments de l'inégalité parmi les hommes* (fig. 11).[99] This sensible Hottentot had come to know European civilization, but preferred to return to his own people. By using this anecdote and using the illustration on the frontispiece Rousseau elevated the Hottentot to the emblem of what was in his eyes man's happiest state of life. It also shows how high the estimation of the Khoikhoi had risen a few years after the publication of Kolb's book. After the Khoikhoi had for a long time been the embodiment of human degradation, they could be used by Rousseau to showcase human bliss in 1755. For this Kolb had paved the way.

for man: "[...] ce période du développement des facultés humaines, tenant un juste milieu entre l'indolence de l'état primitif et la pétulante activité de nôtre amour propre, dut être l'époque la plus heureuse, et la plus durable. Plus on y réflichit, plus on trouve que cet état étoit le moins sujet aux révolutions, le meilleur à l'homme" ([...] the period of the development of the human faculties, maintaining a happy balance between the indolence of the primitive state and the irrepressible activity of egoism, must have been the happiest period, and the longest-lasting. The more one thinks about it, the more one realises that this state was the least prone to revolutions, the best for mankind) (Rousseau, 1964, III, p. 171). For a discussion of Rousseau's borrowings see Fauvelle-Aymar (2002, pp. 270–275).

[98] Kolb (1719, p. 547). Simon van der Stel (1639–1712) was commander of the Cape from 1679 to 1699 and later became governor of the Cape Colony.

[99] After a drawing by Eisen (Rousseau, 1755).

CHAPTER THREE

EXPEDITIONS FROM FORT LIJDSAAMHEIJD: THE VOC AND THE GEOGRAPHY OF SOUTHERN AFRICA IN THE BEGINNING OF THE EIGHTEENTH CENTURY

When Peter Kolb published his book, the VOC had already been exploring southern Africa for half a century. It did this for reasons completely different from Kolb's. Whereas Kolb had an encyclopaedic interest in the exotic world of the Cape and was incensed at the misrepresentation of the Khoikhoi, the VOC was only interested in exploring the economic potential of southern Africa. At the insistence of the few dignitaries of the VOC with scholarly interests, such as Hendrik Adriaan van Rheede tot Drakenstein (1636–1791) and Nicolaas Witsen (1641–1717), scientifically relevant knowledge had occasionally been collected by VOC expeditions at the end of the seventeenth century, but in general economic priorities determined what information was needed. In practice this amounted to a combination of economic, geographic and political intelligence: knowing what a certain region produced and understanding the power relationships and the access routes. The information collected was captured in travel journals and on route maps for use within the network of the VOC. Only in exceptional cases were the travel journals published and could they thus become part of the knowledge about southern Africa that was evaluated by geographers. The conversion of raw data into orderly scientific knowledge was possible in principle because in the eighteenth century no sharp distinction was made yet between knowledge that had been gathered with scholarly objectives in mind and information that had been obtained for other reasons (see chapter 4).[1]

Because of the narrow objectives pursued during the gathering of information, the only specialists deployed in the course of explorations by the VOC in southern Africa were surveyors and persons with some knowledge of geology and botany. Their reporting was more or

---

[1] In making this distinction between "raw" *information* (data) and ordered and interpreted *knowledge* I follow Peter Burke (2000, p. 11).

less professional and, since it was bound to rules, uniform. Compared with the work of scientifically trained travellers, whose field of interest was wider and who were more abreast of the scientific literature, the contributions by servants of the VOC were meagre. But until scientific travellers had developed the ambition and the ability to undertake long journeys into the interior, it was only the VOC that was able to collect information about unknown territory in the interior.[2] This was done intermittently for as long as the VOC governed the Cape.

Within a few years after the establishment of the refreshment station at the Cape, expeditions were already sent into the interior in the hope of recovering some of the huge investments the VOC had made to set up and maintain the refreshment station. Initially it was necessary to determine whether there was gold in the interior. After 1680 the search switched to copper after concrete reports had been received about copper reserves in the north-west of South Africa. Despite moments of scepticism caused by disappointing reports by expeditions, the search for gold continued into the eighteenth century. The demand for gold also prompted the VOC's most ambitious project in Africa of the eighteenth century, namely the establishment of an 'office' (*comptoir*) in Mozambique. The VOC maintained this fortified trading post at Delagoa Bay, known as Fort Lijdsaamheijd in the last years of its existence, from 1721 to 1730 (fig. 12). The fort was situated on the place of the present Maputo (formerly Lourenço Marques), the capital of Mozambique.[3] It was meant to serve as the basis for reconnaissance journeys into the interior, two of which reached present-day Swaziland and the eastern part of South Africa (the province of Mpumalanga).

In the contemporaneous literature this trading post is recorded only in the *Aanmerkelijke Ontmoetingen* (Notable Encounters), the 1744 autobiography of the surveyor Jacob de Bucquoi (1693–1760) who lived there in 1721–1722, and in the *Rampspoedige Reize van het O.I. schip de Naarstigheid* (Fateful Journey of the East Indiaman *Naarstigheid*) of 1761 by Jacob Francken. When the shipwrecked Francken reached Delagoa Bay in 1757, the fact that the VOC ever had a trading post here was already forgotten history, because the shipwrecked sailors were astonished when one of them was addressed in Dutch with the words "goeden dag

[2] Apart from journeys undertaken mainly in the present Mozambique by the Portuguese in the 16th century.

[3] The main studies in this regard are: Coetzee (1948, 1954), Sleigh (1993, pp. 679–716), Paesie (2002, pp. 151–212).

Bootsman" (good day boatswain) by a "very black" coastal resident.[4] The last remains of the voc fort subsequently disappeared beneath Maputo. Although the history of the trading post belongs to the *petite histoire* of the voc, it affords some understanding of the confrontation between generally accepted geographical concepts and the real world. It shows at what cost the voc clung to the highly fictitious picture of the interior of southern Africa that was presented in geographic literature and cartography. More generally, this case shows how representations can sometimes completely obscure reality.

Fort Lijdsaamheijd

Relying on the views of the interior of southern Africa generally held in the seventeenth and eighteenth century, but without much practical knowledge, the voc decided in 1719 to establish this post. It soon proved to be an investment with few prospects. In the first place the fort was situated in a malaria region, which turned it into yet another of the many graveyards of Europeans in the tropics. There were other setbacks as well. In 1722 Fort Lagoa was captured by English pirates, who abducted a number of the personnel (including Jacob de Bucquoi) and destroyed the fort. In 1728 a mutiny broke out in which almost half of the men took part. This mutiny was put down with an iron fist. In 1729 a military force of the fort was massacred by inhabitants of the region. Meanwhile, the commercial activities were minimal. Against the background of this history it was fitting that Fort Lagoa—named after the bay at which the fort was built—was renamed "Fortres de Lijdsaamheijd" (Fortress Patience). Curiously enough, this renaming coincided with a brief period of optimism and enterprise. The new name was given to a strengthened and expanded fort with more cannon and better buildings. Land had been bought in the vicinity that was given new names captured on a map: "de fontaijn strook" (the wells strip), "Capelles dal" (Capelle's valley), "'s Konincks Vlackte" (Koninck's plain; the latter two being called after commanders of the fort), "Noord" and "Zuid Revier" (North and South River).[5] In 1727

[4] Francken (1761, p. 52).

[5] Nationaal Archief (NA), 4.VEL 203–5, "Caert van 's Comp. nuw aengecoght Land" ('map of the Company's newly bought land'), three sheets, dated 8/9–12–1726.

everything still indicated that the VOC had come to stay. Besides the high death rate among the men and other setbacks, however, there was the problem that trading did not come up to expectations in the following years either. In the first years of the fort's existence the main exports were a modest number of copper ingots and ivory, and in later years slaves. But all this was not enough to justify the continued existence of the post. That the VOC had established the post at all and kept it going for as long as ten years was a consequence of the demand for cheap gold and the hope—based on the geographic literature—that this could be found not far from Delagoa Bay.

Gold played an important part in the inter-Asian trade of the VOC, especially in India. Until 1680 it could be had cheaply in Japan and after that in Persia for some years.[6] In the eighteenth century the Persian supply dropped sharply, and this may well have been a reason why the VOC wanted to open up new sources in Africa. However, this is not very clear, since in the "instruction", the document with directions for the future commander of the Delagoa Bay post, gold is mentioned as a possible trading commodity besides others such as ivory, wax and ambergris.[7] The "instruction" of 1719 from the "Gecommitteerde Bewinthebbers" (special Commission of Directors), who had been charged with the policy on Delagoa Bay by the directors of the VOC, to the Council of Policy at the Cape mentions "to discover what benefits could be gained for the Company" once a post had been established at Delagoa Bay.[8] The nature of these benefits was not specified. However, from the context and later correspondence it can be deduced that gold was the main issue after all. De Bucquoi says that among the personnel sent to Delagoa Bay optimistic rumours were doing the rounds about the good prospects for trading in gold before their departure.[9] The location of the establishment also pointed towards gold trading. Although the VOC had bought land at a bay in Terra di Natal in 1690 (the site of the present South African city of Durban) it was not this site that was the first choice for a trading post, but Delagoa Bay, about

[6] Schöffer & Gaastra (1982), Jacobs (2000, pp. 113–115, 122–125, 255).

[7] Cape Archives (CA) C2341, Politieke Raad: Memories en Instruksies, "Instructie voor Sr. Willem van Taak gaande als Opperhooft naar rio de Lagoa en terra di Natal", 29–1–1721, f. 388.

[8] CA C405, Inkomende Briewe, Instruksies "Gecommitteerdens en geauthoriseerdens van d'Oost Ind. Comp. ter Camer Amsterdam" to Governor De Chavonnes, 23–12–1719, f. 176.

[9] De Bucquoi (1744, x).

which the VOC knew less and where it had no property rights at the time. According to the prevailing geographic representations, however, it would be much easier to gain access to the gold-producing interior from Delagoa Bay. One Cape governor called the possibility of trading gold in Delagoa Bay "the soul and the marrow of the enterprise there" in 1724.[10]

The trade policy of the VOC and early modern geography came together in Delagoa Bay. The VOC urgently needed gold for the Asian trade after the supply from Asia had virtually ground to a halt. It was known from the geographic literature that gold would be available at a low price in southern Africa, particularly in the empire of Monomotapa. Since the sixteenth century reports had been doing the rounds that the Portuguese were receiving enormous quantities of gold from Monomotapa. The main source of information in this regard in the geographic literature of the seventeenth and eighteenth century was *Da Asia* (1552) by the Portuguese João de Barros (1496–1570). De Barros describes the court ritual of the Monomotapa, the dignitary after whom the region was called, and mentions the gold mines.[11] Later authors, such as André Thevet (1502–1590) and Filippo Pigafetta (1533–1604), embellished De Barros's story with their own additions. Thevet said that the Monomotapa travelled by elephant and Pigafetta thought that the Amazons lived in Monomotapa.[12]

In the Dutch geographical literature gold from Monomotapa is first mentioned in the *Itinerario* (1595–6) by Jan Huygen van Linschoten (1563–1611). A few years after the establishment of the refreshment station at the Cape (1652), VOC expeditions were therefore already sent into the interior to make contact with Monomotapa. Initially these expeditions were even given Van Linschoten's imaginative map of southern Africa. As these "landtochten" (overland journeys) had no success, they were discontinued after 1664. The lack of success caused an occasional sceptical VOC servant even to doubt the very existence of Monomotapa, which was shown on almost all maps of southern Africa until the beginning of the nineteenth century.[13] But then Monomotapa

[10] CA C1467, Uitgaande Briewe, De Chavonnes to Gecommitteerden, 31–3–1724, f. 409.

[11] Randles (1959, pp. 63–72).

[12] Randles (1959, pp. 75, 81).

[13] See Huigen (1996a, pp. 23–60) and Stone (1995, p. 45). A similar persistent misconception, of more recent, eighteenth-century origin, was the non-existent Kong

is not a total figment of the imagination. It has been proven that a large indigenous state did indeed exist in the border area of present-day Mozambique and Zimbabwe. Partly under Portuguese influence this state disintegrated in the sixteenth century. Gold was also found there, but only in small quantities.[14]

Monomotapa

In the European cartography and geography of the seventeenth and eighteenth century Monomotapa maintained an almost unassailable presence.[15] For example, the tenth part of the *Grande Atlas* (1668) of Johan Blaeu (1596–1673), the voc's house cartographer, contained a map of southern Africa with Monomotapa on it.[16] The essence of the accompanying description goes back to De Barros. The representation of Monomotapa as rich in gold was confirmed by oral sources. For example, a Dutch governor of Mauritius, Isaacq Lamotius, had heard a story about Monomotapa.[17] This report was later given to successive commanders of Fort Lijdsaamheijd as a source of intelligence. According to this report the Portuguese received fist-sized gold nuggets from Monomotapa. They had even had a palace built for the King of Monomotapa close to one of the gold mines:

> The Emperor of Mononigo, who is at the same time also Emperor of Monomotappa [...], has his Portuguese town or gold mine Melinde, where the Portuguese at their expense have built a stately palace for him of stone such as that of the Amsterdam town hall. Which is really beautifully decorated in the European style, with excellent mirrors, superb paintings and rich velvet draperies and carpets all embroidered with gold and silver, and other furniture besides. Whereby they have gained so much favour with him that he has given them permission to trade, live, build cities and a fort in his country, and besides that has given them a very extensive territory which they have benefited from for more than

Mountains in West Africa which appeared on maps of this region until the twentieth century (cf. Bassett & Porter, 1991).

[14] See among others Axelson (1973) and Bhila (1982).

[15] Stone (1995, pp. 32–45).

[16] Koeman (1970).

[17] Lamotius was chief commander of Mauritius from 1677 to 1692.

> a century. This majesty and almost all his rulers and subjected peoples have converted to the Christian faith.[18]

This fairy-tale king, possibly invented along Indian lines, had a daughter as heiress who liked to travel by elephant with an impressive imperial entourage.

Once the Company had gained a foothold at Delagoa Bay the first priority was to establish its presence. Local traders did indeed bring a small quantity of gold to the fort. Jan van de Capelle, a VOC employee at Fort Lijdsaamheijd and the last commander of the fort, interpreted this as confirmation of the current geographic representations of southern Africa in an influential memorandum he handed to the Cape governor De Chavonnes (1654–1724) in August 1723.[19] Van de Capelle presented himself in this memorandum as a person who knew the surroundings of Delagoa Bay and wished to record this knowledge for his superiors. He had heard from an interpreter that there had to be gold in abundance: "In the country around [...] Cennes there is said to be gold in abundance, which is bought by the Portuguese in exchange for coral, textiles, sheets, scissors, knives, gunpowder, lead, tin, even for ready-made clothes, hosiery, shoes etc."[20] It was said that this gold was bartered cheaply by the Portuguese, who used flat-bottomed boats to trade in Cenne. Van de Capelle advised the VOC to send similar vessels as well, so that this interior so rich in gold could be reached by river.

---

[18] "De Keiser van Mononigo die ook tegelyk Keyser van Monomotappa [...] is, houd syn Portugeese dorp of goudmyn Melinde, alwaar de Portugeese op haar koste voor gem. een deftig paleys gebouwt hebben, van sodanigen steen als die van het Amsterdamse stadhuys. En 't welke seer heerlyk door haar op sy Europeaans versiert is, met seer voortreffelyke spiegels uytmunttende schilderye en seer costelyke fluwelen kamers behangsels, en tapyten alle met goud en silwer geborduurt en meer andere meubelen. Waar door sy deselvers gunst soo verre getrocke hebbe, dat hy haar vergunt heeft in syn land te mogen handelen, wonen, stede en 't fortresse bouwen, en daar en boven een groot uitgestreckt landt met ryke goud myne vereert, waar van sy ruym een eeuw 't genot gehad hebben, dees majestyt en meest alle syne vorste en onderhoorige volkeren hebben 't Christen geloof, aangenomen" (CA C733, Inkomende Briewe, "Affricaens Berigt van Isaacq Johan Lamotius", f. 40–44).

[19] CA C2440, Politieke Raad, Verklarings, "Aan Den Wel Edelen gestrengen De Chavonnes en Raad ordinaris van Nederlands India en gouverneur in loco etc. benevens den E. Agtbaren Raad van Politie", Van de Capelle, 3-8-1723, f. 62–88.

[20] "In 't land rondom [...] Cennes soude goud in overvloed wesen, 't geen door de Portugesen werd ingekocht voor corallen, lijwaten, lakenen, scharen, messen, kruyt, lood, thin, jaa selfs voor gemaakte klederen, koussen, schoenen etc." (CA C2440, Politieke Raad, "Aan Den Wel Edelen gestrengen De Chavonnes", f. 66).

A glance at a map showed how easy this would be: the "river St. Esprit according to the map flows more than 200 miles inland, past the city of Monomotapa, and also through the empire of that name where the gold is actually said to occur."[21] As he said himself, the map from which Van de Capelle had acquired his topographic knowledge came from the *Geographie* of the famous seventeenth-century French cartographer Nicolas Sanson (1600–1667). Indeed, on this map, entitled "Pays et Coste des Caffres: Empires de Momotapa Monoemugi", in Sanson's *L'Afrique, en plusieurs cartes novveles* of 1656 one can see that the rivers on the south-eastern African coast extend into the coveted interior (fig. 13).[22] Van de Capelle thinks that one of them, the "Spiritus Sanctus, or as we say St Esprit, is actually at La Goa".[23] On the map the mouth of the river flows past the fort, and upstream it passes along Monomotapa, the "capital city" of the eponymous "empire". If the board of the VOC would send suitable boats, the coveted land of gold could be reached without much effort. Van de Capelle added linguistic proof of the presence of gold as well. The local inhabitants use words that mean gold and silver.[24] If there is a word for it, then the object must also exist.

This was music to the ears of the "Gecommitteerde Bewinthebbers", the commission charged with the policy on Delagoa Bay by the directors of the VOC. They only had to send a few boats to start dealing in gold. In their letter of 13 December 1724 to the Cape governor they confirmed once again, referring to Van de Capelle's memorandum (just to be on the safe side), the current geographic views that strongly correspond to Blaeu's map.[25] Until the last days of the fort and after

---

[21] [De] "'revier St Esprit volgens de caerte wel ruym 200 mylen landwaerd, voorbij de stad Monomotapa, en voorts door het ryk van die naam daer eygentlyk het goud soude vallen, doorloopt" (CA C2440, Politieke Raad, Verklarings, "Aan Den Wel Edelen gestrengen De Chavonnes", f. 68).

[22] Sanson (1656).

[23] "Spiritus St off gelyk wy seggen St Esprit is eygentlyk aen la goa" (CA C2440, Politieke Raad, Verklarings, "Aan Den Wel Edelen gestrengen De Chavonnes", f. 70).

[24] CA C2440, Politieke Raad, Verklarings, "Aan Den Wel Edelen gestrengen De Chavonnes", f. 77.

[25] CA C416, Inkomende Briewe, Gecommiteerden aan gouverneur De Chavonnes, 13–12–1724, f. 21–30. In their letter the Gecommitteerden mention among others "'t Lage Ethiopien" and "'t Meijr Zara", which also occur on Blaeu's map. The Council of Policy passed on this information to the commander in Fort Lagoa (CA C1468, Politieke Raad to Van de Capelle, 8–10–1725, f. 587–8).

even more disappointments, Van de Capelle clung to the traditional geographical representation of the map.[26]

Exploration journeys

Attempts to penetrate the interior by river had already been made shortly after the establishment of the fort. Exploration by boat only served to show that within a short distance from the mouth the rivers were no longer navigable. In any case, it was impossible to see much of the surrounding country from the river: "so far the country is mostly flat, like the dykes in Holland but with very steep banks, without any trees or shrubs but thickly overgrown everywhere with reeds 6 to 8 feet high, so that one cannot see through or over it".[27] The flat-bottomed boats the VOC sent later did not make it possible to penetrate farther into the interior. The interior could only be explored overland.

The VOC had already gained considerable experience with exploratory journeys into the interior of Africa in the seventeenth century. Just a few years after the establishment of the refreshment station at the Cape, expeditions were sent out northward—first to find Monomotapa and later to investigate whether the copper in Namaqualand in the north-west of present-day South Africa could be mined. To make sure that the required information was brought back, the expeditions were put together carefully and given specific instructions. A surveyor was often sent along to collect useful cartographic information. Questions about the nature of the country and people had to be answered in the journal. By the end of the seventeenth century this system of geopolitical data collection had produced a fair amount of information about the north-western part of the present South Africa that was not only circulated within the VOC network, but was also used by geographers such as Olfert Dapper (1636–1689) and François Valentyn (1666–1727)[28] at an

---

26 CA C433, Inkomende Briewe, Van de Capelle to governor De la Fontaine, 2-5-1730, f. 2-134.

27 "[...] het landt [...] tot hier toe is meeste Egaal van hoogte gelijk de dijke in holland dog seer stijl aen de kante, sonder eenige boomen, off boschagie overal dik bewasse met ried van 6 a 8 voete hoog, so dat men niet over nog door mag sien" (CA C733, "Rapport vanden E gesaghebber Jurriaan Cloppenburg [...]", 16-7-1-8-1721, f. 17).

28 Olfert Dapper was a doctor by profession and François Valentyn was a minister of the Calvinist church; the spelling Valentijn is also found.

early stage. The eleven overland journeys undertaken from the Cape to the north-west of South Africa in the seventeenth century ensured an accumulation of geopolitical information. Bit by bit a picture could be constructed, with each next expedition knowing more when it departed than the previous one.[29] This practice was not properly followed during the expeditions from Fort Lijdsaamheijd.

The first overland journey from Delagoa Bay, from 9 to 22 August 1723 to the so-called Blue Mountains, comprised 19 men under the fort's second-in-command Jan Christoffel Steffler and was badly organised. No instruction with guidelines had been drawn up. In addition, the guides they had hoped to engage en route refused to co-operate. Wherever they came the soldiers encountered resistance. The natives advised the expedition not to continue, gave incorrect information, withheld food or simply disappeared when the expedition approached.[30] For example, when after climbing over steep mountains (possibly in the present Swaziland) they arrived at a village, the entire population fled into the mountains, leaving children and animals behind. The local population was scared to death (15 August). On 18 August Steffler committed a serious blunder. In the company of seven other members of the group (including two interpreters from the coast) he had gone looking for a path. They did not find it and wandered about in the bush until they encountered a group of armed local inhabitants, one of whom was the guide they had been promised earlier. However, the guide refused to co-operate, whereupon Steffler put a pistol to his breast. The man consented under duress. But when they crossed a stream, they were attacked. While Steffler was being carried across on the back of a soldier, he got an assegai in the back. Others were also wounded, including Slighting, the commander of the soldiers. The locals withdrew when the soldiers fired at them. Steffler and Slighting died the same night. The survivors returned to the main camp.[31] A report of the events was drawn up later in the fort in which it was concluded that Steffler had taken too many risks.

As usual, the journal of the overland journey was sent to the Cape. Governor De Chavonnes added comments for the directors in

[29] Compare Huigen (1996a, pp. 23–60).

[30] With reference to nineteenth-century expeditions in Africa, the anthropologist Johannes Fabian notes that deceit was the response of Africans to strangers they regarded as dangerous (Fabian, 2000, p. 208).

[31] CA C744, Annexes to Inkomende Briewe Cape Archives Repository, f. 60–8.

Amsterdam. He found that Steffler "had made many observations in 7 to 8 days". However, what worried him was the "inhabitants' concern at the discovery of the nature of their region". Nevertheless, it was necessary to undertake further exploration journeys in order to become familiar with the "roads and location of this region" and to find out what trading possibilities it offered.[32]

The next expedition only left in 1725. This journey was better organised and had more members than before. The company now counted 31 "coppen" (heads): "an assistant, a sergeant, a corporal, a drummer, an assistant surgeon, an ox handler, 20 soldiers and 5 sailors, and 8 persons in number for serving as interpreters and as accompanying boys [i.e. a total of 39 persons—S.H.], and for carrying our provisions 10 beasts of burden and 8 animals for slaughter".[33] The travellers received an instruction from commander Van de Capelle that defined the objectives of the enterprise, something that had been neglected in the preceding overland journeys. The purpose of this expedition was to reach regions such as Inthowelle, where gold would be found. They had to get there with the aid of guides who would be recruited en route. In each region they were to find the tribal chief and inform him that they wanted to trade. Meticulous notes were also to be kept of the environment:

> Furthermore note must also be taken of what the natives, their customs and marital practices are like, this both in the countries mentioned above and in the countries traversed; also the lay of the land, rivers and other geographical features [...] furthermore observing everything closely and keeping a proper daily account of everything that appears worthy of note, so as to hand it over to us in addition to the report of their activities upon return.[34]

---

[32] "[...] veel opmerkinge in de tijd van 7 à 8 dagen [...] syn waargenomen"; "bekommeringh van den Inlander omtrent het ontdekken van de nature van haar landstreek"; "wegen en situatie dier landstreeke" (CA C1467, Uitgaande Briewe, f. 404–9).

[33] "[...] een adsistent, een sergeant, een corporaal, een tamboer, een ondermeester, een leermeester der draagbeesten, 20 soldaten en 5 matroosen, ook als spraakmeesters (tolken) en bij lopende jongens agt in 't getal ende om onse Victualij te voeren 10 draagbeesten en acht slagt beesten" (Punt 1976, p. 44).

[34] "Vervolgens sal ook moeten werden gelet, hoedanigh de naturellen, ommegang en bijwoning [karakter, gewoonten en huwelijksvorm] dier menschen bestaat, en dit sowel in de hier vorengenoemde landen, als die geene door welke den weg sal comen te vallen, ook op de situatie, revieren en verdere gelegentheeden van dien [...] latende wijders omtrent alles een Naukeurig oog gaan, en houdende van al wat aanmerkenswaardig voorkomt, een Nette dagelijxe aanteekening, om het selve by haar terugkomst

However, no consideration was given to the careful recording of geographic information. No surveyor went along, and no instruction was given to record the direction of travel and distance covered. It is therefore difficult to trace the route followed by this group, even if the keeper of the journal, De Cuijper, did his best to keep the notes up to date by making his notes every day using the drummer's drum as a table.[35]

According to the editor of this travel journal, the South African historian Punt, the expedition crossed the Lebombo Mountains north-east of the present South Africa.[36] In my view the few indications given in the travel journal justify this assumption. This much is certain: the party crossed the Incomati river mentioned in the journal on 1 July; on 5 July they crossed a "small mountain range", which could be the Lebombo Mountains, and then continued on 9 July on the South African side of the Lebombo Mountains (not named) between the mountain range and the Incomati river until, on 12 July, they arrived at a place which, according to a local inhabitant, was "one day's travel for a Negro" from the Sabie River (a river in the eastern part of South Africa). After the skirmish on 12 July the company fled back to Mozambique the same night "across horribly large rocky mountains" and enquired the next day "about the Incomati river, which should lie on the right", i.e. towards the south. They crossed the Incomati that same day.[37] What is essential for determining that the expedition had been in South Africa is the fact that they crossed a mountain range both on the way out and the way back—this can only have been the Lebombo Mountains on the border between Mozambique and South Africa—and that between the two crossings they had been one day's journey "for a Negro" from the Sabie River. The expedition must have reached the region of the present Kruger National Park. Nevertheless, it does not seem possible to me to reconstruct the route of the expedition as exactly as Punt did on the basis of the scanty data;[38] the report does not contain sufficient data for this. Distances and points of the compass are not given in the journal. If the details had been recorded with the same accuracy as

benevens 't rapport van hun verrigting aan ons over te geven" (CA C751, Annexes to Inkomende Briewe, 1725, f. 8–12).

[35] "[...] terwijl ik besig was om mijn register te vervolgen sittende te schrijven voor de geweer cap op de Trommel [...]" (Punt, 1976, p. 74).

[36] Punt (1976, p. 90).

[37] Punt (1976, pp. 50, 55, 68).

[38] 38 Punt (1976, pp. 70–72, 90–100).

during the seventeenth-century overland journeys in South Africa, this would indeed have been possible.

The De Cuijper expedition encountered the same problems as the Steffler expedition did. They were sent off into the interior virtually blindfolded. Without maps or navigator they were led by guides from the coastal strip who knew the interior only as far as one day's travel; after that they had to rely on other local guides. Furthermore, the company encountered resistance quite soon. Villagers already fled on the second day, and after crossing the Incomati on 1 July De Cuijper was informed that the local tribal chief intended to murder the travellers. However, before the point of armed confrontation was reached, the locals first employed less violent tactics. The pressure was increased gradually.

The indigenous population knew that the expedition depended on information and guides. Attempts were made to withhold these from the travellers or incorrect information was supplied, for example that no water would be found in the direction the company wished to travel. Sometimes someone was prepared to show the way with or without payment, but such guides did not stay with the expedition for long. On 9 July, for example, a guide had been obtained with great effort on the South African side of the Lebombo Mountains, but he took off that same day when the expedition was surrounded by "some five hundred Negroes". Two other guides also deserted afterwards. Finally, on the 12th, De Cuijper decided that the group should turn back because they did not have enough provisions to reach the lands of gold. Nevertheless, he remained firmly convinced that the land of gold did exist, for in his journal he wrote on the same day: "[the] chief of that people is clad in a golden suit of armour, and all his subjects in copper".[39] As the locals did not yet know that the travellers wanted to return, they resorted that same day to the strongest means of stopping the passage of the expedition—an attack. De Cuijper recounts this moment as follows:

> [...] while I was busy writing my register, sitting writing at the drum in front of the gun awning, I heard my friend [Jan Mona, commander of the soldiers] some twenty paces away from me shouting at the top of his voice: Alarm! Alarm! Shoot! Whereupon I jumped up in a hurry, ran to the gun awning and quickly passed on the flintlocks, and the guard

[39] "[...] [de] opperste van die Landaart [dat volk] gaat gekleed met een gouden harnas, en al sijn onderdanen met Copere" (Punt 1976, p. 68).

> fired and killed a Negro who was coming at us with shield and assegais on the spot. At the same time we saw around us as many Negroes with assegais and shields as there was grass on the ground, making an awful noise by whistling, shouting and blowing their horns, so that we could not hear one another no matter how close, we ordered our people to draw up and close ranks.[40]

A skirmish followed. In the confrontation the company lost its pack-oxen and had to withdraw across the plain to the mountains. During the night the expeditionary force then crossed the "horribly large rocky mountains". On the Mozambique side the problems persisted: again an indigenous army was waiting for them, and when the soldiers had the opportunity to ask why they were being threatened in this way, De Cuijper was told: "because we were passing through their country". While they were crossing the Incomati, they also heard the locals on one side warning those on the other "that they had to keep an eye on us, because we had shot and killed one there".[41]

In the 1960s or 1970s a small monument was erected in the Kruger National Park between the Crocodile Bridge and Lower Sabie rest camps to mark the spot of the confrontation between the expedition and the local population. The historian Punt had identified the place of the event. A bronze plaque on a wooden pole pays homage to the rather unheroic skirmish of 12 July 1725. However, the chances that this took place on the spot where the little monument was erected are very remote, because the scant data make an exact reconstruction impossible. In any case, what was more important to Punt was the first glimpses the European travellers had had of the area of the Kruger National Park a few days before the confrontation. In his words: "5 July 1725 is an important date in the history of the Lowveld and the Game

---

[40] "[...] terwijl ik besig was om mijn register te vervolgen sittende te schrijven voor de geweer cap [afdak voor geweren] op de Trommel, so hoor ik mijn maat [Jan Mona, de bevelhebber van de soldaten] omtrent twintig stappen van mij afgrijselijk schreeuwen alert alert schiet, waar op ik haastelijk op sprong, lopende bij de geweer Cap en gaf de snaphaenen haastelijk aan, waar op de Schildwagt vier [vuur] gaf en schoot een neger ter plaatse dood, die met schild en assegaijen op ons aan quam, op den selven tijd sag men ront om ons so veel negers met assagaijen en schilden als grasen ter aarde, makende een afgrijselijk geluid, door fluijten, schreeuwen en blasen op haar horens, dat men malkander niet kon verstaan hoe digt men was, wij Commandeerden ons Volk te Renjeeren en Sluijten [zich in slagorde op te stellen en de rijen te sluiten]" (Punt 1976, p. 74).

[41] Punt (1976, p. 74).

Reserve, because on that day the first white people saw the area."[42] In the tradition of the colonially tinted, older South African historiography, Punt not only brought back the expedition from the past, but also promptly elevated it to a *lieu de mémoire* of white South Africa. With this in mind the text of the travel journal, entitled *The first Europeans in the Kruger National Park*, was published by Punt on the occasion of the centenary of the Kruger National Park in 1975.[43] Prince Bernhard of the Netherlands even wrote a somewhat rambling foreword in his capacity as chairman of the World Wildlife Fund.

Fiasco

The primary reason for the failure of the overland journeys from Fort Lijdsaamheijd was their relatively inefficient execution. The soldiers who were to undertake the expeditions did not possess the skill to find their way independently in the African bush. They lacked 'bushmanship' and therefore had to rely completely on unwilling guides.[44] The necessary understanding of local political relations fell short in the face of well-considered and coordinated resistance. In a situation where they were dependent on the co-operation of the indigenous population, this shortcoming was perhaps the greatest error the expedition made.[45] The resistance of the indigenous population seems to have been coordinated, as each group warned the other, and well considered because the response was intensified stepwise: first incorrect information was provided, then guides were withheld, then violence was threatened and then armed confrontation followed. While the indigenous population was effectively organised, it seems as if the VOC was constantly improvising and, if resisted, all too ready to resort to brute force.

---

[42] Punt (1976, pp. 92, 94).

[43] The text appeared in 1976 with an introduction and explanation by Punt, and provided with an English translation, by the National Parks Board of Trustees.

[44] Cameron, McLaren and Cooper (1999, p. 337) define 'bushmanship' as "the competence to interact meaningfully with newly encountered environments in the achievement of predetermined goals".

[45] An observation by Johannes Fabian concerning the success (or lack thereof) of scientific travellers of the nineteenth century is applicable here as well: "We have seen in detail and from many different angles that the success of expeditions depended on qualified personnel, sufficient funds, good logistics, and luck. But what counted more than any of these, was politics [...]. European explorers had to be able to identify and deal with power structures" (Fabian, 2000, p. 157).

Moreover, the expeditions did not sufficiently contribute to an accumulation of information.[46] The VOC's tried and tested information gathering methods were not applied (cf. chapter 4). Instead of transferring the information about the interior to maps, transient memories were relied on, which at Fort Lijdsaamheijd all too easily succumbed to malaria or assegais. This also explains why so little of this knowledge can be found in subsequent VOC documents and none of it made its way into the geographic literature.[47]

Fort Lijdsaamheijd was a disastrous undertaking in all respects. A large number of lives were lost and the VOC incurred massive financial losses. According to Sleigh, 490 of the 620 men (about 80%) who were sent to Delagoa Bay over a period of ten years died, most of them from malaria.[48] Fort Lijdsaamheijd was a speculative undertaking based on incorrect geographic knowledge. By 1727, according to the calculations of the VOC, this had already resulted in a loss of 177 049 guilders.[49] However, the policymakers of the VOC cannot be blamed for this miscalculation, since all publications of the seventeenth and eighteenth century painted the same picture of a rich and powerful "empire" of Monomotapa in the interior of southern Africa which was merely difficult to reach. The business acumen foundered on deficient geographic knowledge. In its decision to close down the settlement, the Cape Council of Policy specifically referred to misconceptions about the geography, but these were now attributed exclusively to the personnel of Fort Lijdsaamheijd. The personnel of the fort was not only too lazy to explore the interior thoroughly, it also kept coming up with fairy tales about the riches of Monomotapa which relied on geographic literature such as the *Algemeene Wereld-beschijving* by A. Phérotée de la Croix of 1705 and other 'suspect' sources—which, incidentally, came from the archives of the VOC.[50] The men on the spot were blamed for the fiasco.

---

[46] I borrow this criterion for measuring the degree of success of exploratory journeys from Bruno Latour (1987, pp. 215–257).

[47] That is to say, I have found no references to expeditions from Delagoa Bay in documents concerning journeys to the eastern part of South Africa.

[48] Sleigh (1993, pp. 706–707).

[49] Paesie (2002, p. 204).

[50] Among others, the report by Lamotius on Monomotapa mentioned earlier (CA C85, Resolusies van die Politieke Raad, 11-6-1730, f. 130–9).

CHAPTER FOUR

TRADE AND SCIENCE: REPORTS OF THE VOC EXPEDITION BY HENDRIK HOP FROM 1761–1762

The voc's 'exploration policy' also had to prevent the leaking of information to potential competitors as far as possible. The limited availability of cartographic information about the Cape Colony when it fell into English hands in 1795 bears testimony to the success of this aspect of the policy. Although the Cape Colony at the end of the eighteenth century was the best-mapped territory outside Europe,[1] English colonial officials complained that the voc had done nothing to put the interior of the colony on the map.[2] The numerous maps which were drawn especially at the end of the voc period during the administration of one of the last governors, C.J. van de Graaff (1785–1791), had been shipped to the Netherlands.[3]

None of these maps was published in the eighteenth century. Only a limited number of voc travel journals were printed, with the co-operation of voc officials or other persons at the Cape. François Valentyn published the travel accounts of the journeys by Simon van der Stel (1639–1712) and by the Stellenbosch magistrate Johannes Starrenburg (1677–?) at the beginning of the eighteenth century in *Oud en Nieuw Oost-Indiën* (1724–1726). The governors Simon and Willem van der Stel (1664–1733) had given him the journals. In 1792 an English translation of the journal of a rescue expedition searching for survivors of the Grosvenor, the English ship wrecked in 1782, was published by Captain Edward Riou. Riou had two copies of the travel journal of the rescue expedition: one sent by the brother of the journal-keeper Van Reenen and the other by governor Van de Graaff, who had sent it to Joseph Banks, the secretary of the Royal Society in London. The scientific significance of these published travel journals varied. The journal of the rescue expedition was published mainly for journalistic

---

1 Koeman (1952a).

2 Barrow (1801, p. 8).

3 Koeman (1952a). C.J. van de Graaff (1734–1812) was a military engineer with a special interest in cartography and fortifications.

reasons; the wreck of the Grosvenor had given rise to much publicity and speculation about the fate of the survivors, and the journal of the last rescue expedition, in 1791, was able to prove that none of the survivors of the shipwreck was still alive.[4]

From a scientific point of view the journals of Simon van der Stel's 1685–1686 expedition and of an expedition headed by "burgher-captain" Hendrik Hop (1716–1771) in 1761–1762 are of greater interest. Both travel journals gave an impression of the north-western parts of southern Africa. Van der Stel had travelled up to Namaqualand, and 76 years later Hendrik Hop even crossed the "Great River" (Orange River) into southern Namibia. Of these two travel accounts, the account of Hop's expedition in the eighteenth century received the greatest publicity. The report of Van der Stel's journey was hidden away in *Oud en Nieuw Oost-Indiën*, François Valentyn's opus of five thousand pages on the territory of the VOC in Asia; this was never translated in the eighteenth century, and foreign readers' access to it was therefore limited.[5] The "Dag-verhaal" (daily narrative) of the Hop expedition, drawn up by surveyor Carel Frederik Brink (?–1780) and supplemented with other reports and provided with illustrations, was published by the Amsterdam publisher Schneider in the *Nieuwste en beknopte beschijving van de Kaap der Goede-Hope [...]* (Latest and concise description of the Cape of Good Hope [...]). Schneider published a French translation at the same time entitled *Nouvelle description du Cap de Bonne-Espérance* [...]. A German translation appeared in Leipzig in 1779.[6]

The *Nieuwste en beknopte beschryving* was the first book with new information on the Cape since the appearance of the *Journal historique du voyage fait au Cap de Bonne-Espérance* by the abbé De la Caille in 1763. For this reason alone the *Nieuwste en beknopte beschryving* was already significant. In addition, much of the available knowledge about South Africa was included in it. The first part contained a "brief" summary of the existing literature on South Africa. The "Dag-verhaal" in the second part was provided with notes and 12 illustrations in which zoological knowledge on southern Africa had been collected (fig. 14). The

[4] Huigen (2001, pp. 36–39).

[5] A modern edition of the part about the Cape Colony appeared in the Van Riebeeck series in the seventies (Valentyn, 1971–1973).

[6] *Neue kurzgefasste Beschreibung des Vorgebirges der guten Hoffnung nebst dem Journal eines Landzugs in das Innerste von Afrika durch das Land der grossen und kleinen Namacquas* [...] Leipzig, Weygandsche Buchhandlung (1779). In the text below I will always refer to the Dutch edition, the *Nieuwste en beknopte beschryving* (1778).

"latest" in the title referred, as explained in the preface ("Voorbericht"), especially to this second part, of which the "Dag-verhaal" formed the core. The virtue of the first part was mainly its brevity—it was nothing more than a compilation of information from Peter Kolb's *Capvt Bonae Spei Hodiernvm* and the abbé De la Caille's *Journal Historique.* One year earlier, in 1777, exactly the same had been done in a book with almost the same title, the more comprehensive *Nieuwe algemene beschryving van de Kaap de Goede Hoop* (New general description of the Cape of Good Hope). But the "latest" book of 1778 also contained illustrations and previously unpublished texts with scholarly explanations in the second part. For those who were interested in South Africa, it was therefore a better buy than its 1777 precursor.

At the time of its appearance, the *Nieuwste en beknopte beschryving* of 1778 filled a void in European knowledge about southern Africa, because up to that point nothing had been published about the region north of the Orange River. Yet the "Dag-verhaal" of the surveyor Brink and most of the other documents published in the "second part" of the *Nieuwste en beknopte beschryving* had not been written with scientific objectives in mind.[7] Like other VOC travel journals, the "Dag-verhaal" had to comply with economic and geopolitical needs for information on the South African interior. The editor of the "Dag-verhaal" in the *Nieuwste en beknopte beschryving* lifted the text out of the utilitarian VOC circuit and turned it into a scholarly text by adding annotations and illustrations and, in a certain sense, introducing it with the brief overview of the existing literature on South Africa in the first part ("eerste stuk") of the *Nieuwste en beknopte beschryving.*

In seventeenth and eighteenth-century empirical science it was not unusual to allocate a scientific status to utilitarian knowledge. Robert Boyle (1627–1691), one of the founders of experimental science, referred to the reliability of the narrative of simple, honest merchants. Before Boyle, Montaigne (1533–1592) in his essay on cannibals ("Des Cannibales"; *Essays* I, XXXI) advocated that simple people should be regarded as reliable observers.[8] VOC documents had in the past already

---

[7] In the *Nieuwste en beknopte beschryving* the title "Dag-verhaal" is used for the travel journal of the expedition and as subtitle for all documents, including the travel journal, in the "second part" of the *Nieuwste en beknopte beschryving.* Refer to the annex for the titles of these documents.

[8] Lux & Cook (1998, p. 180); Dear (1985, p. 156); Van Gelder (1997, pp. 103–104).

played a part in scholarly debates. In 1673, for example, Boyle had been willing to revise his views on the origin of ambergris on the basis of a VOC report that had been captured by English pirates. The journal stated that ambergris did not originate from the "excrements" of a whale, but from a tree of which the roots extended into the sea and secreted a resin that rose to the surface. Boyle had to admit that this report was incomplete because it did not discuss the question whether all types of ambergris were formed in this manner. However, the benefit of the report was its credibility, because "it had not been written by a Philosopher to broach a Paradox, or serve an hypothesis."[9] The report had been written by a simple merchant for his principals, purely with the intention of describing a certain fact. According to Boyle, merchants' travel journals were for this reason worth more than the treatises of scholars whose aim was to prove a certain theory and who were inclined to bend the 'facts' to their purposes.[10] Hence the travel journal of Hop's expedition was assured of the interest of the learned world.

I want to examine the double life of the "Dag-verhaal" in more detail below: firstly, how the text originated in 1765–1762 within the VOC network and secondly how the journal was published in 1778, a few years later, within scholarly circles as the latest scientific knowledge on southern Africa. The link between these two histories is an animal that was still unknown at the time and caught the public attention at the Cape in 1760.

"Een hier te lande nog gantsch onbekend Dier"

In 1760 the elephant hunter Jacobus Coetsé Jansz. (1730–1818) was the first European to visit southern Namibia. He returned in November of that year with a tale about a people that wore textiles and had long hair and a "taanachtige" (tan-coloured) skin and with reports of a strange animal.[11] The story about the unknown people with a light skin colour, long hair and dressed in clothes made from textiles was a permanent feature in the discourse of the VOC about the interior of southern Africa. It was also a motif in European geography of southern

[9] Carey (1997, p. 287).
[10] Carey (1997, p. 287).
[11] "Relaas" of Coetsé in Godée Molsbergen (1916, II, pp. 18–22).

Africa. As already mentioned in the previous chapter, a firm conviction prevailed until the beginning of the nineteenth century that the interior of southern Africa was governed by the "empire" of Monomotapa, which was said to have the same level of civilisation as south Asian societies and whose residents therefore also were familiar with textiles. The VOC's quest for trading partners ensured that the story of the long-haired, light-skinned people dressed in textiles regularly surfaced in the Cape VOC documents.[12] What was new in the "Relaas" (narrative) of Coetsé, however, was the report about a strange animal:

> In this land of the great Amaquas [in southern Namibia] a large number of lions and rhinoceroses were also found, as well as another animal still quite unknown here, which is not as heavy as an elephant but, having a much higher body, leads the undersigned to assume for this reason, as well as because of its long neck, its hump and long legs, that if it is not the true camel, it is at least a type of camel; these animals being very languid and slow of movement [...].[13]

Coetsé was able to shoot two animals and to keep a calf alive for two weeks on milk and bran. When it died, he took along the skin. Unfortunately it was not possible to form a picture of the appearance of the adult animals on the basis of the skin.

Coetsé's tale was picked up by an important burgher from Stellenbosch, the captain of the citizen cavalry Hendrik Hop.[14] He approached the Council of Policy in Cape Town with the proposal of an expedition to the land that Coetsé had visited, to the benefit of the Company and the colony. The burghers were to equip an expedition at their own expense and the Company needed to supply only arms, tools and barter goods. The Council of Policy was in favour, and also supplied two specialised personnel: the surveyor Carel Frederik Brink, to draw a map and "dagelijxe aanteekening van al het merkwaerdige te houden" (to keep a daily record of everything worth noting) and the gardener Auge to collect unknown plants.[15] All this was repeated in an Instruction together with further instructions.[16]

---

12 Huigen (1996a, pp. 23–60).
13 Godée Molsbergen (1916, II, p. 21).
14 Mossop (1947, pp. 72–87).
15 CA C139, Resolutions of the Council of Policy of Cape of Good Hope, f. 284–300.
16 This Instruction ("Instructie") was published by Leupe in the nineteenth century (1852).

Reporting techniques

The organisation of the expedition led by Hop revealed new as well as familiar elements. What was new was that the initiative for the expedition had not been taken by the VOC, but by the burghers and also that the personnel of the expedition consisted largely of burghers and their servants—17 "Europeans" and 68 "bastard Hottentots".[17] Only Brink and Auge were employees of the company. The third expert, the surgeon Rijkvoet, who came along to do geological research, was a burgher. Like earlier expeditions, the previous VOC expedition (which was sent to the Eastern Cape in 1752) had still consisted mainly of VOC personnel. Hop's expedition launched a new trend, because after this expedition the VOC made use of burghers for exploratory journeys in most cases.

On the other hand, the techniques that were to ensure the collection of reliable information had been established long before. Since the start of the exploratory expeditions into the interior at the time of Jan van Riebeeck the Council of Policy had sought to discipline travellers to ensure transparency of their reporting. The traveller had to be nothing more than "the hand that collects objects, the eye that observes and describes them",[18] an extension of his principals. In order to ensure this, experts were added to the expedition and an "Instruction" was drawn up that governed the conduct of expedition members, especially of the key persons.[19] Two of the three experts, the gardener and the surgeon, were charged with collecting soil samples and plant specimens. The surveyor had to textualise the information in the form of a route map and a travel journal. The information collected by the expedition had to be put into a format that could be physically transferred, so that back at the Cape it could be interpreted and compared with the results of earlier expeditions.[20] Overall, techniques of this kind were a precondition for successful accumulation of geographic and natural history knowledge in the eighteenth century.[21] At Fort Lijdsaamheijd, this had been neglected, with the result that little was achieved there.

---

[17] Mossop (1947, pp. 14, 16).
[18] Bourguet (1997, p. 177).
[19] Leupe (1852).
[20] Bourguet (1997, p. 173).
[21] Latour (1987, pp. 215–257). Organisations (besides the VOC) with a comparable capacity to collect information over large distances and then to react to it were the Spanish crown and the Jesuit Order (cf. Harris, 1998). Referring to *actor network theory*

Just how firmly the techniques followed were rooted in an established tradition appears from the fact that it was the journal of the expedition headed by Simon van der Stel in 1685–1686, which was by now 75 years old, that was given to Brink as a model.[22] Although this could be interpreted as a sign of conservatism, the Council of Policy at the Cape was actually ahead of its time in the seventeenth century compared with the situation prevailing elsewhere. In other countries the regulation of the conduct of participants in scientific expeditions started only in the eighteenth century.[23]

As in earlier cases, the objectives of the expedition were utilitarian. The long-haired wearers of textile aroused interest as potential trade partners.[24] Ethnography was subservient to economic benefit. The unknown animal might also be economically useful: "it [...] could be possible that these [animals] are used as camels by the local inhabitants and therefore could be of much benefit to this colony".[25] In the end, however, the results proved to be of scientific interest only.

The expedition departed on 16 July 1761 and returned on 27 April 1762. It found no trace of the long-haired, light-coloured, textile-wearing people, and the copper reserves found were unfortunately not viable. However, on 5 October 1761, in the southern part of Namibia, the expedition did see the "first *kameleopardali*" (the unknown animals); they shot a female and captured a calf. From an economic point of view, however, this was no great find: "because of the shortness of the body compared with the other members and its curious gait it is not likely that this animal can be put to any use".[26] In other words: as the giraffes had no adequate back, they were not suitable for use as pack

---

Harris (1998, p. 276) describes this capacity as follows: "the operation of long-distance networks depended fundamentally on their ability to sustain the circulation of people and things from the center to the periphery and back again: in reciprocating patterns of decision and action, intelligence and directive, the knowledge gained through repeated cycles of travel to the periphery strengthened the network and allowed it to extend its reach, project its power, and (in Latour's phrase) 'act at a distance'". See chapter 10 for the use of actor network theory in the research of journeys of exploration.

[22] Leupe (1852, p. 375).

[23] Kury (1998; 2001, pp. 91–146); Harbsmeier (1995, pp. 23–24).

[24] Leupe (1852, pp. 372, 375).

[25] "[...] het [...] soude kunnen sijn, dat dezelve door de inwoonders aldaer als de cameelen selve wierden gebruykt en dus ook van veel nut voor deese colonie soude kunnen sijn" (Leupe 1852, p. 376).

[26] "[...] weegens de kortheijd van 't Lijf, teegens d'andere leeden, in desselfs wonderlyke beweeging, niet te denken, dat dit beest tot eenigen dienst gebruijkt kan worden" (Mossop 1947, pp. 32, 34).

animals. On 16 or 17 October a male was shot and measured up.[27] Brink also made a drawing of the animal, copies of which reached Europe by various routes.[28]

The surveyor Brink kept the journal ("Dag-verhaal") and drew a map of the route followed (figure 15).[29] After Brink's journal had circulated among members of the Council of Policy, the results were discussed at the meeting of 23 November 1762. No comments were made about the journal—that is to say, nothing about it was minuted. The main recommendation was that specialists had to be sent from the Netherlands to research the copper reserves further.[30] The Leiden Cabinet of Natural History (Cabinet voor Natuurlijke Historie) received the skin of the giraffe calf the expedition had captured. In 1770 this specimen was described by the director of the Cabinet, the Leiden professor Jean Sébastien Allamand (1713–1787).[31] The same Allamand was to play an important role a few years later in the publication of the *Nieuwste en beknopte beschryving*, which included Brink's "Dag-verhaal".

---

[27] Mossop (1947, p. 36).

[28] Rookmaaker (1989, p. 291). It is generally assumed that Coetsé's giraffe of 1760 was not the first to be observed in southern Africa. Rookmaaker (among others) assumes that a giraffe was already sighted by a voc-expedition in 1663 200 kilometers south of the Orange River in the present South Africa (Rookmaaker, 1989, p. 15). The keeper of the Journal of this expedition, Van Meerhoff, mentions "twee camelen" (two camels). The assumption that Van Meerhoff meant giraffes depends on the opinion that the word 'kameel' (camel) *always* referred to giraffe in the South African context. This view is supported by the fact that the traveller Wikar in the second half of the eighteenth century speaks of a 'kameel' when he encounters a giraffe and that Coetsé in 1760 draws a comparison with a camel. The term 'kameel' for giraffe is related to the Middle Latin name *camelopardus* for giraffe. However, the assumption that the word 'kameel' can always mean giraffe as well is not supported by the *Woordenboek der Nederlandsche taal* (2000) (lemmas: "kameel", "kameelpaard" and "giraffe") and the *Middelnederlandsch Woordenboek* (1998) (lemma: "cameel"). The meaning of camel for giraffe was evidently not established firmly enough to be included in the main Dutch historical dictionaries. In addition, all other giraffes before 1800 were found north of the Orange River in Namibia, if we ignore the "camelen" of 1663. Rookmaaker therefore suspects an incorrect observation in 1663. It is in any case uncertain whether Van Meerhoff saw a giraffe in 1663. Coetsé's giraffe was the first to draw the attention of the voc; the "camelen" of 1663 were added to the archives unnoticed. See also Foster (2004, p. 32) in this regard.

[29] Koeman (1952b, Pl. IV).

[30] CA C140, Resolusies van die Politieke Raad, f. 354–368.

[31] Rookmaaker (1989, p. 291).

The publication of the "Dag-verhaal"

The second stage in the history of Brink's "Dag-verhaal" was its publication in the *Nieuwste en beknopte beschijving van de Kaap der Goede-Hope* together with other documents that had been drawn up by members of the expedition. In this way information that had been gathered with economic objectives became part of scientific knowledge (compare chapter 3). It is not clear who took the initiative for the publication. The first part ("eerste stuk") of the *Nieuwste en beknopte beschryving* was nothing more than a concise version of Peter Kolb's comprehensive *Capvt Bonae Spei Hodiernvm* of 1719 and De la Caille's *Journal historique* of 1763. The "latest" knowledge was to be found in the second part ("tweede stuk") of the *Nieuwste en beknopte beschryving* entitled "Dag-verhaal gehouden op een landtogt [...]", which is separately numbered and contains, besides the actual "Dag-verhaal" by the surveyor Brink, other documents as well. Briefly, the content of this second part is as follows:

1. "Dag-verhaal" by the surveyor Brink of the expedition of 1761–1762, with annotations by Allamand and Klöckner and 12 illustrations of animals (pp. 1–79).
2. An ethnographic description by burghers Thieleman Roos and Petrus Marais of the population in the regions the expedition had visited (pp. 80–91).
3. The report by the surgeon Rijkvoet on the copper reserves discovered (pp. 92–95).
4. A description of the wax tree ("waschboom") (pp. 96–108).[32]

(See annex 2 for the exact titles of the relevant sections of the second part.)

An informal scholarly network seems to have played an important part in the publication of the four texts in this second part. The indications to this effect are spread over the *Nieuwste en beknopte beschryving*, starting with the Preface ("Voorbericht"), in which the Amsterdam reverend A. Buurt (1748–1781) is thanked for his role in the publication of the "Dag-verhaal": "All we will say as a preface is that for the 'Dag-verhaal' of the overland journey, copied after the original, as it is kept in the archives of the Cape [we have to thank] the Right Reverend and

[32] *Myrica cordifolia*; in Afrikaans: *wasbessie*; in English: *wax myrtle* or *waxberry*.

Very Learned Mr A. Buurt".[33] There are no indications that Buurt ever visited the Cape.[34] It is possible that the same person who made the copy of the "Dag-verhaal" also sent a letter about the "wax tree" to Buurt: the Cape reverend Bode.[35] Bode could have sent copies of the Cape documents to Buurt, who in turn made these documents available to the publisher.[36] The contact between Rev. Buurt and Rev. Bode probably did not have a theological basis. Buurt was not only an eminent reformed dogmatician, but also a famous collector of exotic birds.[37] He may have had contact with the Cape and with Bode in this regard.

Allamand, with some help from the Amsterdam surgeon J.C. Klöckner (1726/7–1778), then raised the scientific quality of the "Dag-verhaal" (at the request of the publisher?) and thereby changed the character of the text by adding notes and illustrations. The result is that especially the simple travel journal written by the surveyor Brink was transformed into a scholarly text.

Allamand's involvement in the editing was an obvious move. In 1749 he became the first professor teaching natural history at the University of Leiden. In addition, he was head of the Leiden Cabinet of Natural History since 1751. His scientific reputation was based especially on the Dutch edition of one of the most important reference works on natural history of the eighteenth century, the *Histoire naturelle* by Buffon (the so-called *Nouvelle édition* that appeared in Amsterdam in 1769–1781). His most important contribution to it was the addition of 41 new descriptions of animal species to Buffon's text. Of these additions the large number of descriptions of animals from southern Africa is noteworthy. Of the 41 additions 25 (63%) were South African. Ten of the 41 descriptions were of animals that had not been described by anyone else yet. Seven of the ten species were also indigenous to South Africa. Allamand's significance as a scientist therefore depended greatly on contacts with South Africa. He obtained his information through an

[33] "Het geen wy alleen voorberichten zullen, is, dat wy het Dag-verhaal van den Landtogt, gecopieerd naar het origineel, zo als het in de Archieven van de Kaap berust [te danken hebben aan] den WelEerwaarden en zeer Geleerden Heer A. Buurt" (*Nieuwste en beknopte beschryving*, 1778, "Voorbericht").

[34] Rookmaaker (1989, p. 37).

[35] Cf. signature of the letter (*Nieuwste en beknopte beschryving*, 1778, p. 98).

[36] For Bode see: SABW (1977, III, p. 77); Engelbrecht (1952, pp. 75–76).

[37] Evenhuis (1974, pp. 68, 70).

informal network of correspondents, of whom the Cape army officer Robert Jacob Gordon was the most important.[38] Allamand became involved in the publication of the *Nieuwste en beknopte beschryving* in his capacity as an expert in the field of South African natural history.

Allamand's annotations to the "Dag-verhaal" in the *Nieuwste en beknopte beschryving* are limited to these fields of competence. According to the Preface he added 40 notes to Brink's "Dag-verhaal", which refer mainly to the geography and zoology of the Cape.[39] He may also have compiled the piece on the "Nieuwe Ontdekking van den Waschboom" (New discovery of the wax tree).[40] The main consideration in this text is in the economic benefit that could be obtained from natural products at the Cape; for example, the berries of the wax tree might be used to produce wax.[41] Towards the end, however, this piece wanders off into zoology, Allamand's field of expertise, with assertions about the pharmaceutical qualities of hyrax urine ("dassenpis") and remarks about whales at the Cape.

The other texts in the second part—the report of surgeon Rijkvoet about copper reserves and the reports of the burghers Thieleman Roos (1728–1780) and Pieter Marais (1726–1775) on ethnography—were published without annotations.[42] These texts were newsworthy enough to be included, but fell beyond Allamand's field of expertise. The report by Roos and Marais, "Aan den wel edelen gestrengen Heer, Ryk Tulbagh" (to the Honourable Ryk Tulbagh) is the most homogeneous text in the second part of the *Nieuwste en beknopte beschryving*. In order to carry out the brief in the expedition's "Instruction" to describe unknown peoples, Roos and Marais offer a systematically compiled description of the peoples in the region visited by the expedition. One after the other the different Namaqua groups are listed (their name, area of residence, character and the etymology of their name), the unknown peoples for which the expedition had to search (but about which nothing was found) and the customs and habits ("Zeeden en Gewoontes") of the Namaquas as a whole.

---

[38] Rookmaaker (1989, pp. 118–128; 1992).

[39] Klöckner was responsible for two notes: 38, on the "Tzeiran" (antelope from Siberia) and 39, on the "groote Gerbo of springrot".

[40] *Nieuwste en beknopte beschryving* (1778, pp. 96–108).

[41] Interest in the commercial potential of exotic plants was common in the eighteenth century (Kury, 2001, pp. 28–34, 40–43).

[42] *Nieuwste en beknopte beschryving* (1778, pp. 92–95 and pp. 80–91 respectively).

The short descriptions of the different Namaqua groups—"Comeinacquas, Tradiamacquas, Cabonas, Korikambis, Keinamacquas"—are presented by Roos and Marais in the order in which the travellers encountered them. People not yet known are geographically located relative to those that are. The report is structured like a route map of the Underground. As the individual descriptions of the peoples are extremely concise, the geographic location of the peoples relative to each other is their most important characteristic. Such a schematic description was sufficient because the relevant Namaqua groups were of no economic interest. By way of supplement, the Namaquas are once more characterised cursorily as a whole at the end. Here a few matters are dealt with briefly which were described more extensively by Kolb, such as their houses, their religion and their wedding and funeral ceremonies. As regards the question—an old ethnographic one in VOC circles and in practice mainly an economic one—whether people were found in the North who were light-coloured, dressed in textiles and resembled the Dutch (pp. 83–4), Roos and Marais regrettably have to reply in the negative. What might be of interest to the VOC is new reports about the "Birinas" who "spoke a language that is quite different from that of the Namaquas" (p. 84) and who trade iron and copper with the informants of Roos and Marais. But unfortunately these are people who rub their bodies with fat and wear animal skins, so that they are no different from peoples who are already known. Commercially speaking, this was not very promising.

The transformation of the "Dag-verhaal" into scientific knowledge

Allamand's most important contribution to the edition consists of the 40 notes to Brink's travel journal. In a limited sense, the notes are an explanation to the text—mainly geographic and ethnographic explanations. These notes contextualise particulars in the text by means of borrowings from the existing literature and cartography.[43] The zoology notes, however, are in the form of short treatises in which the latest knowledge about the animals in question is presented.[44] For Allamand's

[43] Kolb (1719, 1731, 1968) and De la Caille (1763). Notes 2–10, 13–14, 23 on geography in *Nieuwste en beknopte beschryving.*

[44] Notes 1, 11, 15, 18–21, 24–42 on zoology in *Nieuwste en beknopte beschryving.*

treatises in the footnotes, Brink's text serves as a rhetorical anchor in the African world. In three cases (notes 39–41) the zoology notes are even separate from the text. The system of notes leads a life of its own, and on many pages it even elbows the main text off the page.

In the notes Allamand was able to showcase his knowledge of zoology. For example, a reference to the oryx in the "Dag-verhaal" prompts him to present an extensive overview of the different types of antelope found in southern Africa and to showcase the existing knowledge on the not yet so familiar zebra.[45] The question was whether the zebra could indeed be called a donkey. According to an experiment by "Mylord Clive", it could:[46] the latter had brought a female zebra to the England and mated it with a male donkey, after stripes had been painted on the male. This proved the relationship. But, says Allamand, if one had painted stripes on a horse it might also have succeeded, so that the question remained whether the zebra was related exclusively to the donkey.

In a note on "wild horses" (note 27) he speculates about the possibility of the existence of unicorns in southern Africa, of which the Jesuit Jeronimo Lobo (1595?–1678) asserted that they occurred in Ethiopia.[47] According to Allamand, indications of the existence of unicorns in southern Africa had been found. An "attentive observer, who is very trustworthy as well" had written to him that the "Hottentots" had made cave drawings of unicorns.[48] And as the Hottentots were not aware of the debate concerning the existence of unicorns, these drawings would be an objective indication of the existence of these animals.[49]

[45] *Nieuwste en beknopte beschryving* (1778, notes 31 and 37, pp. 46–49, 60–62; note 35; "striped donkey").

[46] Allamand probably refers to Robert Clive (1725–1774), the founder of the British rule over India, who did possess a zebra (e-mail, Kees Rookmaaker, 16 July 2004).

[47] Lobo wrote a report on his experiences in East Africa of which among others a French edition was available in the eighteenth century. His report remained authoritative until the nineteenth century. Allamand, a French-speaking Swiss by birth, used the French translation by abbé Joachim le Grand: *Relation historique d'Abissinie* (Lobo 1728). Lobo mentions among other things that the unicorn runs from forest to forest so quickly that it is difficult to observe it (Lobo 1728, pp. 69–70, 230–231). A recent English translation is Lobo 1984.

[48] Allamand is making an error of ethnographic nomenclature here. The cave drawings in southern Africa were made by the Bushmen (San), not by the Hottentotts (Khoikhoi). It also shows that Allamand's knowledge of ethnography was deficient.

[49] It was not the Hottentots (Khoikhoi), but the Bushmen (San) who made the rock drawings in southern Africa. Allamand's rather unconvincing argument seems to be that the drawings can constitute independent evidence of the existence of unicorns because the Hottentots did not intend to support a party in the debate concerning

An objection, Allamand said, could be that the Hottentots were bad draughtsmen and had actually tried to draw a rhinoceros. Allamand's remarks about the unicorn preceded the scholarly debate about the South African unicorn that would break loose after the publication of Sparrman's account of his travels (first, Swedish edition of 1783), and they indicate the extent to which he was in touch with the latest zoological news from South Africa.[50]

Of course Allamand also provides notes on the giraffe (note 19, 20 and 21), the only animal that is described in Brink's account—other animals are merely mentioned. In one of these notes (note 19) he refers the reader to the illustration: "in order to give an idea of what this peculiar animal looks like, we give a picture of it here" (fig. 14). Twelve illustrations are added to the text, some of which were already used earlier, most of them in editions of Buffon's *Histoire naturelle*, and others newly made for the *Nieuwste en beknopte beschryving*.[51] These illustrations also help to make Brink's utilitarian travel journal more scholarly.

Allamand's notes and illustrations add a discourse that deviates from the main text. The "Dag-verhaal" and the other texts produced by Hop's expedition are geographically organised, in accordance with the conventions of the VOC network. All factual elements are mentioned only in passing and are mainly located spatially relative to one another. The notes and illustrations, on the other hand, draw attention to the facts stated, making use of data from different sources (compare what Allamand observes about the "striped donkey" and "wild horses"). From Allamand's perspective, the facts merit full attention in their own right. The spatial sequence in which they were discovered is less important to him.

Allamand's footnotes and illustrations on the one hand and Brinks's text of the "Dag-verhaal" on the other originate from two different positions of knowledge production. The VOC traveller *casually* accords the collected information a place in the topographic arrangement of his travel account. This information is only important to the VOC as far as as it has economic significance. The armchair scholar, however, comments on the facts in the spatial isolation of his Cabinet.[52] Although

---

the unicorn. I will deal more comprehensively with the debate concerning the South African unicorn at the end of the eighteenth century in chapter 10.

[50] See chapter 10.

[51] Rookmaaker (1989, p. 37).

[52] I will return to this in chapter 10.

the facts stated may sometimes also be of economic value to him—think of the economic potential of the waxberry—he primarily wants to find a place for them within the existing corpus of knowledge. In this way, for example, Allamand links the wild horses of the "Dag-verhaal" to the scholarly debate on the existence of unicorns.

In order to contextualise the "Dag-verhaal" scientifically, Allamand gratefully made use of the possibilities the type page of the book offered him. Because of their scope, the footnotes to the "Dag-verhaal" are not subordinate to the main text; instead, their close connection with the illustrations turns them into the part of the presentation that sometimes draws the greatest attention.[53] They also link up with the first part of the *Nieuwste en beknopte beschryving*, with knowledge compiled from De la Caille and Kolb, which in a certain sense acts as an introduction to the "Dag-verhaal". As a result of Allamand's editorial work, the discourse of the armchair scholar dominates the reporting by the explorers in the *Nieuwste en beknopte beschryving*.

Conclusion

Thanks to their publication in 1778, Brink's "Dag-verhaal" and the other documents the members of Hop's expedition had drawn up were able to make a modest contribution to the discourse on southern Africa. Mentzel borrowed information from the report by Roos and Marais for his description of the indigenous population in the Orange River basin.[54] Mentzel also offers an "Extract" from the "Dag-verhaal" amounting to a transcription of Brink's journal, adding some notes and omitting the part about the journey back.[55] Sparrman and Le Vaillant borrowed geographic data from the "Dag-verhaal", and De Mist used Rykvoet's report in 1802 in the chapter on the mining potential of the Cape Colony in his memorandum on the administration of the Cape Colony for the Batavian government.[56] Until the end of the eighteenth century, the "Dag-verhaal" and the other texts in the second part of the

[53] Cahn (1997, p. 98) describes the footnote as "ein zweites Niveau der Rede" (a second level of discourse). Cahn hereby distances himself from the general view of the footnote as an (often annoying) incidentality.

[54] Mentzel (1787, II, pp. 525–533).

[55] Mentzel (1787, II, pp. 200–230).

[56] Sparrman (1977, II, pp. 103, 120); Forbes (1965, pp. 123–124); De Mist (1920, pp. 67–68).

Nieuwste en beknopte beschryving enjoyed the reputation that they were the product of the longest journey ever undertaken by Europeans to the region north-west of the Cape Colony.[57] According to the preface ("Voorbericht") of the *Nieuwste en beknopte beschryving* it had been a particularly difficult journey, and it would still take several centuries before it was possible to travel to southern Namibia with the same ease as from Amsterdam to Moscow along the postal route. Once these better connections had been established some time in the future, "zal men [...] [echter] geene Oliphanten, Carmelopardalis, enz. op den weg vinden" (one would, [...] [however,] find no elephants, giraffes etc. on the way).

The influence of the second part of the *Nieuwste en beknopte beschryving* on the image of southern Africa nevertheless remained limited, because only a few years later the comprehensive travel accounts of the late eighteenth-century scientific travellers started to appear, starting with the Swedish edition of Anders Sparrman's travel account in 1783. The distribution of these works, especially those of Le Vaillant and Barrow, was much wider than that of the *Nieuwste en beknopte beschryving* (see chapters 6 and 7).

It is especially noteworthy that the texts from the *Nieuwste en beknopte beschryving* operated in two contexts: firstly within the VOC's system of reporting, and after publication as the latest scholarly knowledge on the interior of southern Africa. The editorial additions by Allamand show that the gap between a scholarly discourse and that of VOC travellers had widened since the beginning of the century. The dry journal of magistrate Starrenburg's journey to the north-west of South Africa in 1705 could still be published in the *Nieuwste en beknopte beschryving* by Valentyn in 1724–1726 without any comments. As the scientific literature about the world beyond Europe had become more specialised, this appears to have become more difficult. In 1778 it had clearly become advisable to provide a VOC report with a scholarly commentary. The increased distance between utilitarian travel account and scholarly text is also evident from Georg Forster's disapproving review of Jacob van Reenen's journal of the search for survivors of the shipwrecked English vessel, the Grosvenor, of 1790–1791. A dry summary of events and places visited no longer sufficed, in his view.[58] By the end of the

[57] Thunberg (1986, pp. 322–323).
[58] Forster (1992, p. 363).

eighteenth century the traveller preferably had to be 'philosophical' himself in order to make his account worth while, because such a traveller would know from existing publications what was of interest. If he was able to write with flair, so much the better. Many of the travellers we will meet in the next chapters had this ability.

CHAPTER FIVE

XHOSA AND KHOIKHOI "HOUSEHOLDS": REPRESENTATIONS OF INHABITANTS OF SOUTHERN AFRICA IN THE GORDON ATLAS

Robert Jacob Gordon (1743–1795) is the first of three travellers with a scientific background I wish to discuss in the following chapters. These were people who ventured into the interior with their own research programme, which also implied that utilitarian knowledge was not their main objective. Their research questions were determined in the first place by the contemporaneous scholarly literature. Two of them, François le Vaillant and John Barrow, published accounts of their observations which were widely read; Gordon did not. In his own times he was therefore known only to people who looked him up because of his extensive knowledge of South Africa. At present he enjoys a certain reputation in South African historiography, among other things because of his sympathetic attitude towards indigenous peoples. Gordon appears to have been the only traveller of the eighteenth century who was able to treat people who were completely different from himself as equals, not only the Xhosas of the Eastern Cape, who were much admired by many of his contemporaries, but also the San, whom other travellers usually regarded as trash. A significant part of his work has now been translated into English, and he has een the subject of biographical studies.[1] In the Netherlands and elsewhere he is a relatively unknown figure.[2]

Gordon was born in Doesburg, in the east of the Netherlands, into a military family. Following the family tradition, he became a professional

[1] English translations of the following of Gordon's works are available in English: Raper & Boucher (1988); Smith & Pheiffer (1992); Plug (1995). The main biographical studies of Gordon are those by Barnard (1950) and Cullinan (1992). The Department of Archeology of the University of Cape Town maintains a website where texts by Gordon—especially his travel accounts—are provided together with a translation into English (http://web.uct.ac.za/depts/age/people/Gordon/frameset.htm).

[2] The historical zoologist Kees Rookmaaker in particular devoted attention to Gordon in the Netherlands (especially Rookmaaker, 1981 and 1989). Gordon has become somewhat better known in the Netherlands following an exhibition of drawings from the Gordon Atlas in the Rijksmuseum in 2002.

soldier at an early age. Besides his career, Gordon also took an interest in natural history. For this reason he studied at the University of Harderwijk and undertook a journey to South Africa in 1773–1774. During his stay of thirteen months he undertook a journey into the interior, about which nothing more is known.

Back in the Netherlands he could already be regarded as a specialist on South Africa after his first journey. It is also in this capacity that he is mentioned by Diderot (1713–1784) in the latter's *Voyage en Hollande* (published posthumously in 1817). Diderot called him a young Englishman. This was probably because of the uniform of the Scots brigade Gordon was wearing. As he had a distant Scots ancestor, Gordon and his father had been assigned to the Scots brigade of the Army of the Republic (figure 16). During their meeting, Diderot questioned Gordon about a subject that occupied the scholarly community until well into the nineteenth century: the extended labia of the 'Hottentot Venus' (cf. Chapter 1). Gordon was able to tell him that this fleshy appendage was indeed part of the labia and not an "apron" hanging from the stomach.[3]

In 1777 Gordon left for South Africa for good. Here he became captain in the VOC army and in time rose to the rank of colonel and commander of the Cape garrison. His departure for South Africa at least created the impression that he was driven by an interest in natural history. This can be inferred from a letter to the Hendrik Fagel (1706–1790), a high-ranking bureaucrat in the Dutch Republic and the man who had probably helped him to settle at the Cape.[4] In this letter Gordon defends himself against the accusations of some people that he had gone to the Cape on a whim, in order to roam the country.[5] Le Vaillant provides a similar indication of the nature of Gordon's motives. From the Xhosas Le Vaillant heard a story about "the only decent man of my race" who also, like Le Vaillant, travelled about in order to satisfy his scientific curiosity ("voyageait par curiosité"). He immediately realised that this referred to Gordon.[6]

---

[3] For a discussion of this meeting see Porter (1991, p. 82) and Cullinan (1992, pp. 22–24).

[4] During the reign of the stadhouders Willem IV and Willem V the greffier Hendrik Fagel was a particularly influential political figure in the Netherlands (*Nieuw Nederlandsch Biografisch Woordenboek, III* (1914, col. 390–391).

[5] Nationaal Archief, The Hague (NA), Fagel-Archief, no. 2515.

[6] Le Vaillant (1790, p. 331).

Gordon's life came to a tragic end. In 1795 a British fleet appeared in False Bay with a letter from the deposed stadholder Willem V of Orange (1748–1806), who had fled to England after the French had invaded the Dutch Republic in January 1795. In this letter all Dutch administrators of the Cape were instructed to submit to the authority of the English, who would then administer the Cape on behalf of the stadholder until the latter was returned to office. The VOC officials were happy to oblige. Governor Sluysken (1736–1799) and the military commander Gordon were both Orangists, with an ingrained aversion to the revolutionary Patriots who had taken over the government in the Netherlands and also had sympathisers in the Cape. Nevertheless, in the end Sluysken decided not to transfer authority, as the sovereignty over the colonies did not rest with the stadholder, but with the States General of the Dutch Republic. Following this decision, a half-hearted defence of the Cape was put up. The commander of the armed forces, who because of his years of service with the Scots brigade wrote English fluently (as appears from his writings, which was unusual for Dutchmen in the eighteenth century) and may also have felt some affection for Britain, excelled especially by the crafty withdrawal of his troops from strategic positions. The surrender was signed after a few weeks. Gordon was suspected of treason.[7]

Some time after the surrender Gordon's family heard a shot in the garden. Gordon was found dead. According to one of his officers, Lieutenant P.W. Marnitz 1751–1821), who favoured the Patriots, a bullet had penetrated Gordon's head from the lower right-hand side. According to Marnitz, this ballistic data proved that Gordon had committed suicide, even though the fiscal determined that it was a case of murder. Marnitz added in a footnote that the fiscal had said this at the insistence of the English commander in order to spare the family.[8] Unfortunately, I was unable to find a report of an autopsy in the Cape archives.[9]

[7] Barnard (1950, pp. 406–436).

[8] P.W. Marnitz, "Verhaal van de Overgaave van de Kaap de Goede Hoop aan de Engelschen door Een vriend der waarheid aldaar. In 't tweede jaar der Batavische Vrijheid" (Cape Archives (CA), VC 75, fol. 252–256). It must be noted that the author was a political opponent of Gordon's, as appears from the Batavian calendar used by Marnitz.

[9] I was able to find a testament, however. It was drawn up on 27 June 1780, shortly after his marriage to Susanna Margaretha Nicolet, long before his death (CA, MOOC 7/1/40, no. 81).

Gordon's stay in South Africa produced scientific writings, drawings and maps. Originally these papers formed part of a larger collection that included (stuffed) animals. According to visitors, Gordon's house in Cape Town was an important museum, in which a stuffed giraffe was probably the most impressive exhibit.[10] After his death in 1795 his scientific estate became private English property. In 1913 the maps and drawings, known as the Gordon Atlas, were acquired by the Rijksprentenkabinet. In 1979 Harry Oppenheimer, the South African mining magnate, bought the manuscripts known as the Gordon Papers and made them part of his private library, the opulent Brenthurst Library in the affluent Johannesburg suburb of Houghton. The drawings and writings cover a wide range of subjects, especially botany, zoology, meteorology, geography and ethnography. In this chapter only a small part of Gordon's legacy will be examined—the drawings of indigenous inhabitants of South Africa in the Gordon Atlas. I will examine these representations together with Gordon's writings.[11] Other drawings from the Atlas which have ethnographic significance, such as a set of four copies of rock drawings and drawings of animals and plants which were important to the indigenous population, will be left unconsidered.

Gordon's epistemology

Gordon's scholarly career in South Africa displays the typical characteristics which the historian of science Basalla ascribes to colonial scholars during the first phase of the practice of science in colonies. Gordon was a keen collector of particulars of the foreign world, which were gratefully used and published by the subject specialists in Europe.[12] The results of Gordon's zoological research reached the European public via the Dutch natural historians Vosmaer (1720–1799) and Allamand.[13] Although he was unable to publish anything arising from his work,

[10] Rookmaaker (1989, p. 64).

[11] In this investigation I am referring to the most important ethnographic drawings from the Gordon Atlas: the ethnographic drawings on the large map of South Africa (G3), a landscape showing an indigenous settlement (G32, "Gesigt op het land [...]") and the purely ethnographic drawings (G69–93; except for G92–93 these drawings depict indigenous inhabitants of South Africa; G92 and G93 show a knife and a grave).

[12] Basalla (1967, pp. 611–622).

[13] Rookmaaker (1989, pp. 118–128). For the botanical work see Dyer (s.a., pp. 44–62).

contributions by Gordon to the field of zoology were included in the Amsterdam edition of the *Histoire Naturelle* by Georges Louis Leclerc de Buffon (1707–1788)—the "nouvelle edition" (1766–1771)—that was edited by Allamand. Gordon supplied descriptions which Allamand included as supplements.[14] Nevertheless, Gordon did not see himself as inferior to researchers who did appear in print. For example, among his writings there are critical annotations to the publications of the scientific travellers De la Caille and Sparrman.[15] He also took a critical view of Buffon's more speculative theories; he calls Buffon one of the greatest minds on earth, but also one who had not seen everything with his own eyes.[16] This was a kind way of saying that Buffon, head of the Jardin du Roi, the botanical garden of the French king in Paris, sometimes acted as an armchair scholar.

Gordon was a fundamentalist empiricist, who rejected speculative theorising and regarded scientific research especially as the gathering of reliable facts.[17] In his research practice he followed Buffon closely, for whom the detailed description of natural phenomena was the main task of the study of nature. Gordon's epistemology can also be understood best if we relate it to Buffon's. Buffon rejected Linnaeus' project to describe nature within a taxonomic system on the basis of randomly selected characteristics. According to Buffon, such a taxonomy was not rooted in reality at all and therefore of no value.[18] In contrast, he advocated a meticulous study of nature. Natural science had to collect faithful descriptions of natural phenomena by means of accurate observation. On the basis of these factual descriptions general conclusions could be drawn which in turn formed the

---

[14] Rookmaaker (1989, pp. 123–128); see also chapter 4 of this book.

[15] CA, VC 595, 80–104. Gordon's writings, known as the *Gordon Papers*, are kept in the Brenthurst Library in Johannesburg. For my research I made use of photocopies of the *Gordon Papers* which are kept in the Cape Archives in Cape Town (CA, VC 592–7).

[16] "[...] je regardait Monsieur Buffon, comme ayant été un des plus grands hommes de la Terre, mais il n'a pus voire tout de ses propres ijeux" (CA, VC 595, 148). Buffon had formulated the vastness of the field of research of natural history: "l'Histoire naturelle, prise dans toute son étendue, est une histoire immense; elle embrasse tous les objets que nous présente l' univers" (Buffon, 1749, I, p. 3) (Natural history, in its widest sense, is a vast history; it encompasses all objects presented to us by the universe).

[17] The formulation of hypotheses and theories were disputed subjects in the new, post-Aristotelian natural science (cf. Shapin & Schaffer, 1985, pp. 49–51; Shapiro, 2000, pp. 143–160).

[18] Roger (1963); Eddy (1994).

basis for constructing explanatory theories that had to be empirically demonstrable. In Buffon's words:

> [...] In the study of natural history one must confine oneself to producing accurate descriptions and one must be in possession of hard facts. One must in the first place pursue truth, the essential goal, as we remarked earlier. But thereafter one must endeavour to achieve something greater and occupy oneself with something of greater value, namely the combining of observations, generalisation based on facts, establishing of relations on the basis of analogies. One must strive to achieve such a high level of knowledge that one can determine how particular consequences depend on more general consequences, so that one can compare Nature with itself in its grand operations and so that one ultimately opens up avenues to make perfect the different parts of Physics. A vast memory, meticulousness and attention suffice to reach the first goal. But here something more is needed: a general survey, a steady eye and a mind more trained by reflection than by study. Lastly, one must possess the mental faculties that allow us to discern distant relations, to bring them together and, after having determined the exact measure of truth therein and having weighed the degree of probability, to develop a body of reasoned theories.[19]

Buffon's epistemology comes close to the empirical cycle of the twentieth-century standard model of scientific practice: observation, generalisation and formulation of theory.[20] In Gordon's epistemology we find the first two steps of Buffon's model, i.e. observation and generalisation. The third step, the formulation of theory, is absent. Gordon's faith in science was rooted solidly in observation. He distrusted the formulation of theories, even the theories of Buffon, whom he admired, if in his view they had no empirical basis. This mistrust of theories arose from

---

[19] "[...] dans l'étude de l'Histoire Naturelle on doive se borner uniquement à faire des descriptions exactes et à 's assurer seulement des faits particuliers, c'est à la vérité, et comme nous l'avons dit, le but essentiel qu'on doit se proposer d'abord; mais il faut tâcher de s'élever à quelque chose de plus grand et plus digne encore de nous occuper, c'est de combiner les observations, de généraliser les faits, de lier ensemble par la force des analogies, et de tâcher d'arriver à ce haut degré de connoissance où nous pouvons juger que les effets particuliers dépendent d'effets plus généraux, ou nous pouvons comparer la Nature avec elle même dans ses grandes opérations, et d'où nous pouvons enfin nous ouvrir des routes pour perfectionner les différentes parties de la Physique. Une grande mémoire, de l'assiduité et de l'attention suffisent pour arriver au premier but; mais il faut ici quelque chose de plus, il faut des vûes générales, un coup d'oeil ferme et un raissonnement formé plus encore par la réflexion que par l'étude; il faut enfin cette qualité d'esprit qui nous fait saisir les rapports éloignez, les rassembler et en former un corps d'idées raissonnées, après en avoir apprécié au juste les vrai-semblances et en avoir pesé les probabilités" (quoted from Roger, 1963, pp. 531–532).

[20] Koningsveld (1982, pp. 65–73).

doubt about human cognitive ability. In this regard, he wrote the following to his patron Fagel: "ik ben gansch niet sistematieq, volkomen overtuigt van de swakheid onser vermogens om de werken van den groten Schepper nategaan" (I am not at all systematic, [being] completely convinced of the weakness of our abilities to trace the works of the great Creator).[21] In practice, Gordon occupied himself mainly with the meticulous observation of phenomena.[22]

When making generalisations he operated strictly inductively. He generally ventured to make statements with general validity only after having collected sufficient facts. For instance, he asserted on the basis of altitude readings taken with a barometer that the interior of South Africa was a plateau—which, as it turned out, was a correct assumption.[23]

In the course of his South African travels Gordon also collected information on different peoples in southern Africa. According to his own classification—a generalisation, once again correct, that he based entirely on his own observation—it was possible to distinguish two completely different peoples in southern Africa: the "Caffers" (Kaffirs) who called themselves "Kosaas" (now Xhosas) and the "Hottentots" (Khoikhoi). According to Gordon, Caffers[24] or Xhosas differed from the Khoikhoi in that they spoke a different language, practised agriculture and were in all ways more civilised than the Khoikhoi. The two groups also differed in terms of appearance and "character".[25] He subsequently added the remark that the hunting "Bushmen" (San) and the stock-farming Khoikhoi "in all probability" formed a single group, like the Xhosas in the south-east and the "Moetoaanaas" or "Briquas" (Tswanas) in the north-west of South Africa.[26] Although he distinguishes two main groups, in descriptions and drawings he particularly attempts to present

[21] NA, Fagel-Archief, no. 2533, 24 April 1779.

[22] Michel Foucault's definition of the observing gaze is applicable to Gordon's way of observing: "The observing gaze manifests its virtues only in a double silence: the relative silence of theories, imaginings, and whatever serves as an obstacle to the sensible immediate; and the absolute silence of all language that is anterior to that of the visible" (Foucault, 2003, p. 132).

[23] NA, Fagel-Archief, no. 2533, 24 April 1779.

[24] In the remainder of the text I will use Gordon's ethnonyms with adapted spelling. Gordon's orthography was not consistent.

[25] NA, Fagel-Archief, nos. 2515, 2533.

[26] Note on the drawing of "Moetoaanaas Huijshouding" on Gordon's large map of South Africa (Gordon Atlas, G3).

a picture of the exterior and the culture of specific subgroups, such as the Namaquas (a Khoi group) or the San (fig. 17).

The relatively high degree of naturalism of some of the ethnographic drawings in the Gordon Atlas is exceptional. They were made by someone who wanted to document the appearance of the indigenous inhabitants of South Africa as accurately as possible.[27] Although the figures are mostly stiff and poorly proportioned (fig. 18), an attempt was at the same time made to represent characteristic features instead of reducing them to schematic figures as was customary in, for example, illustrations in published travel accounts of South Africa in the eighteenth century.

Such figures are for example found in Peter Kolb's *Capvt Bonae Spei Hodiernum*, the most important work available at the end of the eighteenth century. Gordon regularly refers critically to Kolb. The numerous ethnographic illustrations in Kolb's book differ in the various editions and are probably not based on sketches, but on Kolb's descriptions of the manners and material culture of the Khoikhoi. Illustrations which seem to be actually based on realistic sketches were a rarity in the literature on South Africa. One example was a representation of the Namaquas living in the north-western part of present-day South Africa in Tachard's *Voyage de Siam* of 1686.[28] Another feature that the ethnographic illustrations in published travel accounts have in common is that they remained limited to illustrations of Khoikhoi and did not yet include representations of the Xhosas of the eastern border of the Cape Colony. An empirically minded researcher such as Gordon was therefore not very impressed with the available ethnographic images of the Cape. He regarded his own observations as so much proof of the unreliability of the existing literature. He made the following remark to his patron, Hendrik Fagel: "As far as the Hottentots are concerned, this must convince someone who reads the latest descriptions of the

---

[27] Documentary drawings are made for the purpose of capturing what is observed (Smith, 1992, p. 53).

[28] The picture of Namaquas in this book is based on a sketch by the Cape pharmacist Hendrik Claudius (see Huigen, 1996a, pp. 51–52). The opposite also happened. Andries Beeckman, who drew pictures of Khoikhoi at the Cape in 1657–1658, based a drawing of a male Khoikhoi on an illustration from Thomas Herbert, *Some Yeares Travels Into Africa & Asia* (De Groot, 2005, p. 24). The National Library in Cape Town has a collection of naturalistic drawings of Khoikhoi by a Dutch artist from the late 17th century published in Smith, 1993.

Cape even more of the shortcomings of those books and make him reluctant to read travel accounts."[29]

The referential meaning of the drawings is important to my argument. Unlike several influential works on the representation of strange worlds and foreign people, such as Edward Said's *Orientalism*, I do not want to leave the empirical dimension of the drawings unconsidered. In postcolonial theory it is generally assumed that representations constitute their own reality and merely create the impression of being a representation of the real world; there is nothing more than an "effet de réel".[30] They are considered to be part of a discourse turned inward. Every Western representation would therefore only be an expression of exoticism, where (representations of) foreign objects and people are given new meanings by the culture that produces these representations.[31] According to these assumptions the empirical dimension of representations would be completely irrelevant.[32]

However, this line of thought overlooks the meaning that eighteenth-century scientific representations had in their own time. Scientific representations of the (foreign) world were intended and were received as attempts to reflect this world as accurately as possible.[33] They relied on a correspondence theory of truth. For example, Gordon attempted

[29] NA, Fagel-Archief, no. 2533, fol. 6–7, 24 April 1779.

[30] Barthes (1968, pp. 84–89).

[31] I follow the definition of exotism given by Peter Mason: "The exotic is produced by a process of decontextualisation: taken from a setting elsewhere [. . .], it is transferred to a different setting, or recontextualized. It is not the "original" geographic or cultural contexts which are valued, but the suitability of the objects in question to assume new meanings in a new context" (Mason, 1998, p. 3). The advantage of Mason's definition of exoticism is that the concept can be used as a special form of the representation or the use of strange objects, where decontextualisation takes place.

[32] Books in which this view is taken are, among others: E.W. Said, *Orientalism* (1978); C.L. Miller, *Blank Darkness. Africanist discourse in French* (1985); P. Mason, *Deconstructing America* (1990); P. Mason, *Infelicities. Representations of the exotic* (1998). Much criticism has been levelled at this approach to the representation of strange worlds, especially at Said's *Orientalism* (1978). However, this criticism has remained limited to issues of discourse theory—the possibility of a critical, Western 'counter-discourse' besides a colonial discourse, where Said (1978) sees only a suppressing colonial discourse; the disputability of the omnipresence of 'strong othering' (extremely negative representation of the Other) in Western representations, as Said contends; the existence of a variety of colonial discourses instead of a Western discourse. To my knowledge the way in which the experience of the strange world undermines the discourse about it has received little attention, however. There is no room to delve more deeply into this problem area here; for a more extensive but certainly not exhaustive discussion see Huigen (1996a, pp. 23–59).

[33] Stafford (1984, pp. 31–58).

to correct the earlier descriptions of southern Africa, as appears from lists of questions he drew up following his reading of Peter Kolb's description of South Africa. In the field he compared assertions made in Kolb's book with his own observations. To Gordon Kolb's representations could only be true if these representations matched the facts. If this was not the case, Gordon's own representations were intended to provide a better correspondence to the real world.[34] This empiricism of the travelling researchers led to an extremely critical view of predecessors' work. Consequently the discourse of the time was to a large degree a collection of unstable hypotheses, which were constantly being corrected at the factual level.

Of course, the ideal of an empirically faithful representation that met eighteenth-century standards was often not achieved in practice. Although close observation and faithful representation were an epistemological ideal the explorers steadfastly pursued, established representational schemes still exerted their influence. Goethe (1749–1832), for example, had to make strenuous efforts in 1786 during a journey through the Alps not to see the mountains as a picture, that is to say within the framework of the existing pictorial tradition based on generalisation. Capturing the specific characteristics of the landscape demanded concentrated attention.[35] We will also see that despite the pursuit of naturalism the ethnographic drawings in the Gordon Atlas nevertheless display traces of traditional schemata from European ethnographic art. This is why many drawings are representations of anonymous, ethnographic types serving as a hook to hang ethnographic features on. Individuals in drawings are often representative of the population group to which they belong, with typical clothes and physical peculiarities. A characteristic of this convention is that people are drawn full size.[36] On the other hand the Gordon Atlas contains drawings intended as portraits, with the name of the sitter indicated. This is unusual, because in the eighteenth century ethnographic portraits of indigenous people were still rare.[37]

---

[34] Gordon regularly makes corrective references to Peter Kolb (*Capvt Bonae Spei Hodiernvm*), for example, in his writings. For the list of questions Gordon compiled see: Smith & Pheiffer (1992, pp. 31–37). The philosophical intricacies of the correspondence theory of truth fall outside the scope of this chapter; see for a discussion of the problem Künne (2003, pp. 93–174).

[35] Stafford (1984, p. 46).

[36] Joppien & Smith (1985, I, pp. 6–9).

[37] Smith (1992, p. 100).

Within the Gordon Atlas as a whole, however, ethnographic drawings are in the minority. There are far more drawings of plants and animals.[38] This does not mean that Gordon was less interested in the indigenous population than in the natural environment, something other scientific travellers of the eighteenth century are reproached with.[39] The indigenous population receives wide attention in Gordon's writings. The relatively small number of ethnographic drawings is probably due to practical impediments: dead animals and plants are quite simply easier to depict than living people being subjected to an unfamiliar experience.[40]

Draughtsman and patron

Although Gordon was the author of the documents in the Gordon Papers, he did not make all the drawings in the Gordon Atlas. Like some other scientific travellers, Gordon took along a "painter" on his explorations, the German VOC soldier Johannes Schumacher (Gordon calls him Schoemaker in his travel journal). Schumacher, of whom little is known, is occasionally mentioned by Gordon in his travel journals. Gordon regarded him primarily as a servant and frequently a source of irritation, because Schumacher became unmanageable if he did not receive enough alcohol on the way.[41] Gordon made drawings himself as well. As the drawings in the Gordon Atlas are usually not signed, it is

---

[38] According to Rookmaaker (1981, pp. 123–124) the Gordon Atlas consists of 16 maps, 52 landscapes, 25 pictures of indigenous people (not counting the illustrations on map G3 and G32), 114 mammals, 110 birds, 30 other animals and 108 plants. This classification does not take into account that some drawings fall within more than one category. For example, G32 shows a landscape, a San settlement and the way in which San slaughter cattle to make peace ("vrede slagten" (slaughter peace))—caption by Gordon). Even if boundary cases are taken into consideration, the Atlas still is of a high natural historical quality.

[39] Pratt (1992).

[40] Smith (1992, pp. 75, 77); Joppien & Smith (1985, I, p. 47).

[41] Cullinan (1992, pp. 32, 56–57, 73, 104–105, 124, 149). Besides his work as a draughtsman, little is known of Johannes Schumacher; only that he arrived in South Africa in 1770, served in Gordon's company in 1778–1781 and worked as assistant in the Arsenal in Cape Town from 1782 to 1789 (*Suid-Afrikaanse Biografiese Woordeboek, IV*, 1968–87, p. 579). The drawings Schumacher made for Hendrik Swellengrebel are kept in the Swellengrebel Archives in St. Maarten (North Holland). Hallema (1951) published the landscape drawings from this collection as a book. The etnographic drawings were used as illustrations in the publication of a fragment from Swellengrebel's travel journal (Hallema, 1932, pp. 131–137).

often not easy to determine who made a specific drawing. The historical zoologist Rookmaaker assumes that Gordon and Schumacher can be said to have co-operated in the production of zoological drawings and ascribes a dominant role to Gordon.[42] In my view, it is possible to draw a somewhat clearer distinction between principal and draughtsman in the ethnographic drawings. I rank Schumacher as an ethnographic draughtsman higher than Gordon, because it is certain that it was Schumacher who made at least five drawings of indigenous people in the Gordon Atlas. Only one ethnographic drawing can be ascribed to Gordon with certainty. However, Schumacher was a draughtsman employed as a servant, as executor of his master Gordon's ethnographic and iconographic programme, which also had consequences for the way in which the drawings were made.

First, Schumacher's role as a draughtsman of indigenous people. Four drawings in the Gordon Atlas can be ascribed to Schumacher with certainty because they are copies of drawings he had made in 1776 for Hendrik Swellengrebel (1774–1803), another Dutch traveller in South Africa, who had undertaken a journey to the Xhosas on the eastern boundary of the Cape Colony one year ahead of Gordon.[43] Schumacher made a total of six drawings of indigenous people of the eastern border region of the Cape Colony for Swellengrebel.[44] Four of these so closely resemble drawings in the Gordon Atlas that the latter can only be copies of the drawings Schumacher made for Swellengrebel.[45] Furthermore, it is sure that Schumacher drew the portrait of the Xhosa chief Coba (fig. 18),[46] as Gordon recorded in his travel journal that he instructed his draughtsman to make a portrait of himself and of Coba and one

[42] Rookmaaker (1989, p. 77).

[43] Hendrik Swellengrebel was the son of a former Cape governor. In 1776–1777 he undertook three journeys through the interior of South Africa (*Suid-Afrikaanse Biografiese Woordeboek, III*, p. 798); Schutte (1982, pp. 3–18).

[44] St. Maarten, Swellengrebel Archives, F1–24–29.

[45] The following drawings in the Swellengrebel Archives (F numbers) and the Gordon Atlas (G numbers) respectively correspond: F1–25 ("Een Caffer Captein" (a Xhosa chief))—G71 ("Caffer"); F1–27 ("Cafferin" (Xhosa woman))—G77 ("Caffer"; in the caption Gordon's subject has changed sex!); F1–28 ("Palo Kaffer")—G72 ("Caffer"); F1–29 ("Mahotti-Kaboo-Polo-Caffer")—G70 ("Mahotti Caffer Captein").

[46] Gordon Atlas, G75. Caption: "Dese Kaffers hieten sig selven Cosaas of Amma-Cosa. Coba Caffer Captein, ses voet lang en git swart met sijn Schild en Assagaijen" (these Kaffirs call themselves Coosas or AmmaCosaas. Coba Xhosa chief, 6 foot tall and pitch black, with his shield and spears); caption at top right: "ik had hem die muts vereerd" (I gave him that cap) (referring to the grenadier's cap Coba is wearing in the drawing).

of his wives during his stay with Coba.[47] The portrait of Gordon was intended for Coba. Schumacher's portrait of Coba has been preserved in the Gordon Atlas together with a portrait of a Xhosa chief's wife (one of Coba's wives?) and Pusega, Coba's daughter.[48]

Of only one drawing in the Gordon Atlas can it be said with certainty that Gordon had a hand in it, and that is the drawing of a settlement of the so-called Strandlopers (beachcombers) in the north-west of present-day South Africa.[49] He mentions in his travel journal that he drew the huts of these people.[50] However, this does not imply that he also drew the human figures, because many of the more complex drawings that illustrate the lifestyle of a subgroup seem to have been made 'at home', in Cape Town, making use of other drawings, as we shall see. It is possible that Gordon contributed only a sketch of the huts. But it was indeed Gordon who drew the sketch of the body of a Khoi woman that is found in his papers.[51] The limited data therefore suggest that Schumacher played a relatively greater role as an ethnographic draughtsman.

If the ethnographic drawings in the Swellengrebel collection are compared with those in the Gordon Atlas, Gordon's role as principal becomes more apparent. The figures in the Swellengrebel collection are intended to represent ethnographic types, anonymous figures whose purpose is to display the external features that are characteristic of the group: "Een Caffersvrouw (met kind)" (a Xhosa woman (with child)), "Cafferin (met kind op den rug)" (Xhosa woman with child (on her back), "Een Kaffers Captein" (a Xhosa chief), "Cafferin" (Xhosa woman), "Palo Kaffer" (Palo Xhosa).[52] The running figure of the "Palo Kaffer", for example, also shows what Swellengrebel mentions in his travel journal (figure 19):

> They wore a karos (cloak) over one shoulder which was held by a strap across the breast and passing in some cases over and in other cases under the shoulder and which hung down their back to just above the calf. The

---

[47] CA, VC 592, 88, 11 December 1777.

[48] "Kaffers Capteins vrouw a 177~~6~~<7>" (Xhosa chief's wife, with date corrected by Gordon; Gordon Atlas, G69); "Pusega Kaffers dogter van kaptein Coba git swart" (Pusega, Xhosa daughter of chief Coba, pitch black) (G74).

[49] Gordon Atlas, G91.

[50] CA, VC 592, 176, 22 August 1779.

[51] "Hottentottin" (Hottentot woman) (Brenthurst Library, MS.107/12; CA, VC 595, 54); see below.

[52] Swellengrebel Archives, F1–24–28.

> karosses, with the hair still on it, seemed to be very well tanned and seemed to be cattle, lion and tiger (meaning: leopard) or other skins. They cover their genitals with a flap of skin which, although long, is nevertheless narrow and hangs loose between the legs, so that at the slightest movement they are exposed [...][53]

The individual figures in the Gordon Atlas—leaving aside the representations of groups for the moment—are in the first place ethnographic types, as there is a striking uniformity in the manner of representation. Almost all figures are shown as frontal views. This has its equivalent in the zoological drawings, where—following the practice in Buffon's *Histoire Naturelle*—Gordon showed animals from the side. According to Rookmaaker, this was done to make it easier to compare the animals represented.[54] The frontal representation of the people may have served a similar purpose.

In Gordon's drawings, however, the frontal view is a 'degree zero' of representation. In the first place, the attitude of the figures can vary. They may be standing at attention or, slightly more elegantly, in the style of established European conventions with a 'Renaissance elbow' at the side, like the chief Coba. Here the European convention of depicting male, often militaristic rulers with a hand resting on the hip has been applied to an indigenous tribal chief (fig. 18).[55] Where many of the martial European examples hold a standard or a field marshal baton, chief Coba holds a bundle of assegais (javelins) in his other hand. On his head Coba wears a Dutch grenadier's cap Gordon had given him. The parallelism between the copper Dutch Lion on the cap with the arrows in each claw (clearly visible in the original drawing) and Coba's bundle of assegais is noteworthy. This parallelism could suggest a degree of subservience to the authority of the VOC—nothing more than this, because the VOC's entire border policy was aimed at keeping the Xhosas, to which Coba belonged, outside the borders of the colony.

---

[53] "Op den eenen schouder droegen zij een caros [mantel], die bij sommigen over en bij anderen onder den schouder met een riem over de borst was vastgemaakt en hen langs den rug tot even boven de kuit hing. De karossen scheenen met haar er aan zeer wel bereid te zijn en schenen van runderbeesten, leeuwen en tijger [bedoeld wordt een luipaard] of ander vellen. Hunne Genitalia bedekken ze door een lap vel, hetgeen wel lang, doch smal en los tusschen de beenen hangt, zodat die door de minste beweeging bloot hangen" [...] Hallema (1932, p. 132).

[54] Oral communication by Kees Rookmaaker.

[55] The 'Renaissance elbow' was also used in other drawings (Gordon Atlas, G88, 77, 79, 83). For the 'Renaissance elbow' in European painting see Spicer (1991). With thanks to Herman Roodenburg, who pointed this out to me.

Many other figures also have European attributes. Chief Ruiter, for example, wears a copper 'breastpiece' (a small copper plate worn around the neck by eighteenth-century members of the military).[56] Several figures are holding a so-called chief's baton ("kapiteinsstaf"), a staff with a copper knob the VOC gave to Khoi chiefs as a symbol of subservience to the authority of the VOC.[57] It is in fact remarkable how many drawings of separate figures in the Gordon Atlas represent tribal chiefs. Their names are stated on the drawings. This greater individualisation of the figures depicted in the Gordon Atlas than in the Swellengrebel collection is related to the greater effort Gordon made to get to know indigenous people.

Hendrik Swellengrebel did not succeed in entering into discussions with the indigenous population. While collecting ethnographic information, he was hindered too much by language problems and was therefore only able to record what he saw, as he indicates in his travel journal. In order to be able to speak to the Xhosas, he needed three to four interpreters because nobody understood both the language of the Xhosas and Dutch, besides which "it is already difficult to convey one's ideas to such a savage people".[58] This is why there is only one portrait of a tribal chief in Swellengrebel's collection, and even then Gordon states that the name in the caption to the drawing is incorrect.[59] Unlike Swellengrebel, Gordon took great pains to build up an understanding with the Xhosas. In his journal he writes how he sang them Dutch songs, learnt their language and names and danced with them.[60] This close contact resulted among other things in the first extensive Xhosa word lists.[61] He even became friends with chief Coba. Before his departure, Gordon instructed Schumacher to make portraits of himself, Coba and one of Coba's wives.[62] The good relation remained intact even after Gordon's departure; it was probably the reason why during a subsequent visit in 1778 in the company of governor Joachim van

[56] Gordon Atlas, G80, 82.
[57] Gordon Atlas (G78, 79, 83, 86). See: Elphick (1985, pp. 191–192).
[58] Hallema (1932, p. 135).
[59] "Mahotti-Kaboo-Polo-Caffer", Swellengrebel Archives, F1–29; CA, VC 592, 90, 12 December 1777.
[60] CA, VC 592, 82, 5 December 1777.
[61] Especially in: "Collection de mots des divers langages de l'Afrique Meridionale", CA, VC 594, 106–127.
[62] CA, VC 592, 88, 11 December 1777.

Plettenberg (1739–1793) Gordon was able to restore the peace on the eastern border of the Cape Colony. Coba trusted Gordon.[63]

The drawing of portraits generally remained limited to the depiction of characteristic features of an individual and recording the name of the person shown in the drawing. However, the Gordon Atlas contains two related attempts to show characteristic facial features as well: the two group portraits of the Namaqua chief Noebè (called "Wildschut" by the Dutch) and the members of his clan, a group of Namaquas in the north-western part of South Africa.[64] Such a rendering of the physiognomy was exceptional in the ethnographic art of the eighteenth century (fig. 20).

Once again it is evident from Gordon's manuscripts that he became thoroughly acquainted with Noebè's clan. During his stay with this group he compiled word lists and described their funeral ritual.[65] This ritual is described in two drawings, one of which depicts the body of Noebè's father, chief Pluto (or Plato)[66] wrapped in his cloak before his burial (see fig. 20).[67] One of the figures is holding a rope which, according to the caption and Gordon's description, is intended to be used to tie up the corpse in the cloak. Noebè (on the right in the drawing with the chief's baton given to him by the VOC) is rubbing his eyes, which corresponds to the remark in Gordon's description that Noebè was very upset at the death of his father, who, according to information Gordon had found in the secretariat of the VOC in Cape Town, was more than a hundred years old when he died. Two figures, on the left in the drawing, are holding up a hand in what could be a gesture of blessing if European conventions were followed in the representation.[68]

---

[63] On their second encounter Coba did not immediately recognise Gordon: "hy bleef in het begin leggen en scheen soubsonneurs [wantrouwig], dog toen hem vroeg of hy zyn broer gordon niet meer kende sprong hy vrolyk op en gaf my vriendelyk de hand" (at first he remained lying down and appeared suspicious, but when I asked him if he no longer knew his brother Gordon he sprang up joyfully and cheerfully gave me his hand (CA, VC 593, 150, 15 October 1778)).

[64] Gordon Atlas, G78 (caption: "Familie der Kleine Namaquas" (Family of the Little Namaquas); names added: "Kareés, 'Kakaûmaap, 'Neigaap, 'Eigaab, Kasaap, Capt. Noebé of Wildschut"); G86 ("Dode Hottentot, So als men hem op de Regter Sijde legt, en in een bind in Sijn Vel of Caross om te begraaven") (dead Hottentot, as he is laid on his right side, and wrapped up in his skin or karos to be buried).

[65] CA, VC 593, 99–101.

[66] In *Resolusies van de Politieke Raad* (1981, IX, pp. 256–257, 13 March 1739) reference is made to a Namaqua chief named "Plato". This could be the same man as "Pluto".

[67] Gordon Atlas, G86.

[68] Compare Bulwer (1974, pp. 109–117).

Another figure shows a tumulus.[69] Again, this drawing illustrates one of Gordon's observations. The drawing can be interpreted as a sequel to the drawing in which the corpse is wrapped in the cloak, because in the same report Gordon states that rocks are piled on top of the grave to prevent wild animals from devouring the corpse. It was also customary to put the bow of the deceased on the grave together with broken arrows (fig. 21).

Gordon's interest in the funeral rites of the Namaquas was roused by his reading of earlier publications on South Africa. Among his papers there is a list of questions prompted by Peter Kolb's comprehensive description of the Cape.[70] Based on this work, Gordon also noted a few matters (with answers) related to the Hottentots' view of death and their funeral rites.[71] Incidentally, descriptions of funeral practices constituted a regular part of ethnographic treatises since the sixteenth century.[72]

Gordon's travels served especially to collect reliable data and to provide answers to questions he had encountered in the literature on South Africa, among others. One of the oldest and probably most curious ethnographic problems was that of the elongated labia of female Hottentots. This question had been a regular topos in the literature on the Cape for a long time.[73] During his first journey after settling in South Africa, in 1777–1778, Gordon examined Hottentot women on several occasions and measured their labia, which indicates that this was a problem about which he urgently required certainty.[74] In one of his examinations he also made a drawing of the so-called Hottentot apron, which was preserved among his papers.[75] Gordon was not the only

[69] "Hottentots Capteins Graf" (Gordon Atlas, G93).

[70] *Capvt Bonae Spei Hodiernum* [...] (Kolb, 1719).

[71] Smith & Pheiffer (1992, pp. 31, 35).

[72] Hodgen (1964, p. 176).

[73] See the large number of references in the register of Raven-Hart's anthology of passages from South African travel accounts (1971, II, p. 494).

[74] He examined Khoi women on the following dates: 25 October 1777, 15 November 1777, 7 January 1778. The so-called Hottentot apron—the *labia minora* that extend beyond the *labia maiora*—is no figment of the imagination, but was observed among San women in the 1970s (Singer, 1978, pp. 124–125). Incidentally, Gordon also examined men for the phenomenon of semicastration—monorchy—where one testicle was removed (cf. Chapter 2).

[75] The drawing of a Hottentot woman (CA, VC 595, 54; Gordon Papers, MS. 107/12). The caption reads as follows: "lengte der uithangende nimphen die, met de labia, als ene continuatie makende 6? duim rynlands [ongeveer 17 cm], in dese vrouw lang. men segt dat zy enige hordes het voor een schoonheid houden, en het langer

one who drew this apron. His contemporary Le Vaillant also produced some drawings of it, one of which was published (fig. 3).

Different drawings in the Gordon Atlas illustrate other ethnographic facts: the use of tails as a handkerchief[76] or spoon,[77] the ox as a beast of burden,[78] the way in which the San "slaughtered peace",[79] and the dances of the Namaquas to the accompaniment of wind instruments.[80] Three drawings depict the disproportionate accumulation of fat on the posterior of San women (staetopygia) by showing that a child could stand on its buttocks.[81] Each drawing of indigenous inhabitants displays the external characteristics of a particular group.

Synoptic (re)constructions

The ethnographic drawings in the Gordon Atlas reflect Gordon's research programme. Most of the drawings represent facts observed, frequently with the aim of confirming or repudiating the European discourse on South Africa. However, a number of the ethnographic drawings have a wider range. They are synoptic (re)constructions of the way of life of a certain group: three of them are separate drawings[82] and five are illustrations on Gordon's large map of South Africa, a monumental work covering two pages of several square metres on which Gordon summarised his geographic and some of his zoological and ethnographic knowledge.[83] These representations of the lifestyle of indigenous population groups offer more than the documentation

maken, door er swaarte aan te hangen. in kinderen vind men even, de traces, van dese singuliere formatie der delen." (Length of the exposed pendant nymphae which, together with the labia, as an extension are 6? Rijnland inches [about 17 cm] long in this woman. It is said that some tribes esteem it a thing of beauty and lengthen them by attaching a weight to them. In children one also finds the traces of this curious shape of the parts.)

[76] Gordon Atlas, G87.

[77] Gordon Atlas, G90.

[78] Gordon Atlas, G83.

[79] Gordon Atlas, G32.

[80] Gordon Atlas, G89.

[81] Gordon Atlas, G3 ("Bosjesman Huijshouding" (Bushman household), G90, 32. *Staetopygia* was an intriguing subject for travellers. For a medical description of this phenomenon see: Singer (1978).

[82] Gordon Atlas, G73, 90, 91.

[83] Gordon Atlas, G3. For comments on this map see Forbes (1965, pp. 111–113). Forbes calls this map "the culmination of Dutch cartography at the Cape" (Forbes, 1965, pp. 111–113).

of an ethnographic fact. At the conceptual level they correspond to Gordon's generalising remarks on the composition of the indigenous population of South Africa. But the drawings follow a classification at a lower taxonomic level. Whereas in his letters and in the caption to the large map (on the drawing "Moetoaanaas Huijshouding") he divides the population into two main groups—the Hottentots and Bushman on the one hand and the Kaffirs (Xhosas) on the other—the drawings in question present an image of the culture and economy of specific subgroups, of their 'households', as Gordon calls them on the large map.[84] The drawings show *in situ* what the members of the group look like, how they make their living and how they live.[85] Traces of intercultural contact such as those that can be seen in portraits of individuals (parts of European uniforms, VOC staff, etc.) are absent in these reconstructions. The cultural group depicted is shown in its reconstructed original state. In some drawings music and dance are also represented, a classical subject in ethnographic art (fig. 22).

These images were made at a different time than the drawings of individual figures. The individual figures were probably drawn from life, as appears from the reference in Gordon's travel journal to the portrait made of Coba. The drawings of "households", however, are idealised typical constructions produced in the studio. They illustrate the appearance of a certain group of foreign people and display the knowledge that has been collected about their culture. The human figures in these representations serve the greater purpose, namely the image of the "household". They are shown performing various actions—dancing, hunting, metal working, tilling the soil—against the background of their huts and their cattle. As a result, the figures in these drawings are more schematic. The aim of the drawing is to reflect collective peculiarities and actions, not individual characteristics.

The genealogy of some drawings of households also indicates that they were not drawn from life. With one exception, which will be discussed later, all representations of people on the large map of South

[84] According to the *Woordenboek der Nederlandsche taal op CD-Rom* (2000), the term 'huishouding' (household) means, among other things, 'society'. The more limited meaning of 'family life' which, according to Van den Boogaart, the term 'huishouding/res familiaris' has in Van Linschoten is not applicable to Gordon (cf. Van den Boogaart, 2000, p. 6).

[85] *In situ* representations of foreign cultures remain popular in the ethnographic museums, for example in dioramas. The intention is to enable the public to experience the foreign cultures as they 'really' are (cf Kirshenblatt-Gimblett, 1991, pp. 388–390).

Africa (Gordon Atlas, G3; figure 23) are based on previous drawings.[86] In their turn, some of the drawings used also have a genealogy. This can be seen in the reuse of figures taken from older drawings. For example, the drawing of Chief Coba's kraal is populated with figures derived from two drawings of a Xhosa man and a Xhosa woman. One of them even harks back to the drawing originally made for Hendrik Swellengrebel by Schumacher. In the representation of the household of the Kaffirs (Xhosas) these individual figures were multiplied and shown in different postures.[87] This drawing of the Xhosa settlement was subsequently reused in slightly modified form on Gordon's large map of South Africa.

One drawing on the large map of South Africa, the representation of the Moetoanas (most probably the Tswanas) has no antecedent (fig. 24). Departing from Gordon's strict empirical epistemology in terms of which drawings are visualisations of observed facts, this drawing presents a picture of a group of which Gordon never visited a settlement. During his visit to the region of the Orange River, he met only a few individual Moetoanas. The drawing of the settlement is therefore a hypothetical construction based on incomplete information he had gathered about the Moetoanas.

Gordon would have liked to obtain more information about these Moetoanas, whom he initially—following the customary colonial usage—refers to mostly as Briquas. On the basis of language and appearance he had rightly classified the Moetoanas he had encountered at the Orange River in 1779 north of the Cape Colony as related to the Xhosas living east of the Cape Colony.[88] His desire to get better

[86] The Xhosa household of the large map is based on Gordon Atlas, G73, and the illustration of the Strandlopers on the map strongly resembles G91. G85 and G89 are incorporated in the representation of the Great Namaquas Household. Elements from G32 ("Gesigt op het land [...]" (View of the country...))—the huts—and G90 ("Nog een schets van het bosjesmans huishouding" (Another sketch of the Bushman household)—women and children—were taken over in the representation of the Bushman household on Gordon's map.

[87] Gordon Atlas, G73, making use of the figures G69 and G71. The drawing of a "Caffer" (G71) multiplied in G73 in turn is a copy of "Een Kaffers Captein" (F1–25) kept in the Swellengrebel Archives.

[88] "[...] sag hier twe briquas [Moetoana's], synde de ene een regte Caffer, so als lang gedagt heb [...] hunne taal is deselve van de Cosaas [of Kaffers], dog een verschillend dialect" (CA, VC 593, 64, 5 november 1779). ([...] saw two Briquas [Moeoeanas] here, one being a genuine Kaffir, as I long thought [...] their language is the same as that of the Xhosas, but a different dialect). This is an instance of Gordon's cautious inductive theorising.

acquainted with them had been aroused by the Moetoanas' industrial products. In the course of his journey he bought a worked hide that was used as a cloak, and the Gordon Atlas also contains a drawing of a knife shown from three sides.[89] He also enquired about houses, the working of iron and agriculture. Some of this information was used in the hypothetical representation of the Moetoana household. A man wearing a cloak such as described by Gordon in his travel journal is placed prominently in the middle of the drawing with his back towards the spectator.[90] The short trousers worn by the men to cover their genitals are also typical.[91] On the left iron is being smelted, which relates to Gordon's interest in the working of iron by the Moetoanas.[92] The way in which iron is being smelted as well as the rest of the drawing, however, appear to be closely based on the assumption that the Moetoanas belong to the same main group as the Xhosas, as is also explained in Gordon's caption to the drawing. However, that the way of life of the unknown Moetoanas was more or less similar to that of these Xhosas—that they therefore had a fairly large huts, tilled the soil, raised stock and worked their iron in the same way as the Xhosas by blowing air into the fire through the horns of cattle, as Gordon had observed during his visit to chief Coba—was nothing more than an assumption based on observed linguistic and external similarities.[93]

---

[89] CA, VC 593, 63, 3 November 1779; Gordon Atlas, G92.

[90] "de briquas of moetjoana dragen een kleinder vel van een calf of jong beest onder de het [sic] grote mantel vel van een harte beest daar de staart in de nek afhangt het vel als een mantel over de schouwders op de borst vast gemaakt, het twede vel hagt meest agter in het grote, niet lager als de dyen, (de mantel hangt tot de hielen [...]" (the Briquas or Moetjoanas wear a smaller cloak of a calf or young cattle underneath the large cloak of a hartebeest with the tail hanging down the neck and the skin over the shoulders fastened at the breast; the second skin usually hangs behind inside the large one, not lower than the thighs (the cloak reaches down to the heels [...]) (CA, VC 593, 64, 5 November 1779); see also the caption to the drawing to the Moetoana household on Gordon's map of South Africa (Gordon Atlas, G3).

[91] "[...] hunne gansche schamelheid, is in een balsak van een bok of calf het hair na buiten" (their entire genitals are in a scrotum of an antelope or calf with the hair on the outside) (CA, VC 593, 64, 5 November 1779); see also the caption to "Moetoaanaas Huijshouding" (Gordon Atlas, G3).

[92] Gordon was unable to obtain much information about the working of iron by the Moetoanas, but there is a picture of a knife "dewelke de moetsjoäna en Capii's Natie seer kunstig maken" (which the Moetoanas and Capii's nation make very skilfully) in the Gordon Atlas (G92).

[93] Gordon records the following about the working of iron by the Kaffirs: "zag by de kraal komende capitein Coba die nog by deze kraal zyn bezoek bleef houden, besig met twe a drie van zyn volk assagayen te smeden, hunne hamer was van yser en hun aambeeld een klip, een zyner soldaten had een ledere zak aan iedere hand

The principles Gordon followed in the construction of this representation can be deduced from his writings. The main principle is the determination that the Moetoanas he had seen in the north resembled the Xhosas who lived far removed from the Moetoanas, on the eastern frontier of the Cape Colony, as regards appearance and language. The reasoning must have been that if they resembled the Xhosas in appearance and language and therefore belonged to the same main group, their settlements and economy also had to resemble those of the Xhosas. This was supplemented with information specific to the Moetoanas, such as their short trousers and cloaks and the fact that they kept goats.

Conclusion

In his ethnography Gordon distinguishes between more civilised and less civilised, which in practice is a matter of whether the group represented corresponds more or less to the Western European standard.[94] The Moetoanas and the Xhosas ("Caffers") are deemed more civilised than the other groups because their lifestyle approximates that of the Dutch in the Cape Colony more closely: they process skins in the same way, are able to forge iron and practise agriculture.[95] In Gordon's case, evaluations of this kind are limited to statements concerning the technical capacity of a culture. No remarks are made on the moral nature of the group described. Neither do they ever lead to 'othering' in the sense of metaphorically positioning them at a distance.[96]

Moreover, Gordon puts his classification into perspective by regarding certain ways of life as determined by ecological factors. For example, he

---

aan wiers voorste eind een beeste hoorn opengemaakt was die in het vuur lag, so dat hy de zakken die by zyne handen open waren toedrukken het vuur sterker aanblies" (when reaching the kraal, I saw chief Coba, who was still staying at this kraal, busy with two or three of his people forging assegais [spears]; their hammer was made of iron and their anvil a stone; one of his soldiers had a leather bag in each hand at the front end of which a cow's horn had been opened that lay in the fire, so that he could squeeze shut the bags, which were open in his hands, to blow up the fire (CA, VC 592, 87, 10 December 1777).

[94] Cf. Huigen (1996a, p. 138).

[95] Remark on agriculture in caption to "Moetoaanaas Huijshouding" (Gordon Atlas, G3).

[96] Fabian (1990, pp. 753–772).

explains the primitive economy of the hunter-gatherers as the result of scarcity of water. Gordon suspects that drought in large parts of Africa makes it impossible to raise stock. The ecology determines whether hunter-gatherers or stockholders live in a given region.[97] Gordon's socio-economic taxonomy is not related to his moral judgement either. He regularly expresses his appreciation of the indigenous population of South Africa. Overall, he finds that at bottom people do not differ much from each other:

> [I] find much greater differences between animals of which we are certain that they are descended from one pair than between people. It strikes someone who travels among different wild folks that he finds them to be the same *human beings* [Gordon's emphasis], although at first sight it seems otherwise.[98]

Reading such a remark makes it difficult to maintain in this instance the cynicism about European representations of foreign people that prevails in postcolonial studies.[99]

Gordon's ethnographic representations form part of an encyclopaedic, scientific project to describe all possible aspects of the South African reality. He takes as much interest in geography, animals and plants as in people. One could say that he is driven by a levelling 'mapping impulse' in which all aspects of the foreign world are equally important and must be mapped literally and figuratively.[100] The large map in the Gordon Atlas (G3) is the culmination of this impulse. The most important cultural groups are represented on the map alongside the most characteristic animals. For the sake of completeness there are even representations of a Cape farm and the anchorage in Table Bay. An

[97] Remark on "Moetoaanaas Huijshouding" (Gordon Atlas, G3).

[98] "[ik] vinde veel meer onderscheid tussen dieren dewelke wy seker weten van een paar te stammen, dan tusschen menschen. het frappeert iemant dewelke onder verschillende wilden reisd, deselve *mensch* te vinden, al schynt het in den eersten opslag anders." NA, Fagel-Archief, no. 2533: the emphasis on 'people' is Gordon's. Georg Forster held a similar view, judging by a statement in his *Ansichten vom Niederrhein*, 1790: "[...] wir bedenken nie, wie ähnlich wir den Wilden sind, und geben diesen Namen sehr uneigentlich allem, was in einem anderen Welttheile nicht Parisisch gekleidet ist" (we never consider how close we are to the savages, and we give all those in another part of the world wrongly that name who are not dressed after the Parisian fashion) (Forster, s.d., p. 731).

[99] As mentioned earlier, Le Vaillant (1790, p. 331) also heard positive comments about Gordon during his journey. The Xhosas in the east regarded Gordon as the only decent white person they knew.

[100] Alpers (1989, pp. 119–168).

overview of the entire South African world is provided, albeit—because of the size of the map—at more than a single glance.

The indigenous inhabitants were amazed at Gordon's obsession with describing and drawing pictures (or having pictures drawn) of everything, but according to Gordon's notes they seem to have taken no offence at his curiosity. The Xhosas were amused at his recording of their words and names. The San in the north-west were surprised at his eagerness to draw everything, but found it a harmless eccentricity:

> [...] nothing could equal the surprise of the Bushmen and the Einiquas upon seeing the drawing. They said we were curious people, but that they now realised that I *coeroe-o-o*'d everything, that is to imitate and to write, and that they now saw that this was why I was wandering so far this way and which they did not understand before.[101]

During Gordon's lifetime these drawings and notes formed part of a larger collection that regularly attracted visitors. Gordon had set up a small museum in his house with stuffed animals, ethnographic artefacts, maps and drawings. He seems to have considered publication, but because of his early demise nothing ever came of it.[102]

Although Gordon's zoological and botanical work was disseminated through the publications of others,[103] his ethnographic research hardly became known outside the circle of people he was acquainted or corresponded with. None of his notes in this field were published. As far as I am aware only two ethnographic drawings of Gordon's collection were used as illustrations in the travel journal of the English traveller William Paterson (1755–1810), who accompanied Gordon on two of his journeys. One is an adaptation of a drawing of a Bushman settlement from the Gordon Atlas. The representation of the Strandlopers in William Paterson's book is clearly a copy of the drawing in the Gordon Atlas (fig. 25).

It was not only Patterson who acquired ethnographic drawings from Gordon. In a letter to his patron Fagel, Gordon asked him to forward a drawing of a Hottentot woman—judging by an earlier reference in

[101] "[...] niets kon de surprise van de bosjesmans en einiquas evenaaren, op het sien der tekening. Sy seiden dat wy wonderlyke menschen waren, dat sy nu sagen dat ik alles coeroe-o-o, dit is voor namaken en schryven, en dat sy nu sagen dat ik daarom so ver heen en weer liep en keek, daar sy te voren geen begrip onder sig van hadden." (CA, VC 593, 14 October 1779.)

[102] Cullinan (1992, p. 165).

[103] See note 14.

the letter and the caption, probably his own drawing of the torso of a Hottentot woman—to the professor of anatomy in Leiden.[104] There are also three ethnographic drawings in the British Library suspected to have come from Gordon's collection.[105] Of one—a Bushman—there is not even an original in the Gordon Atlas,[106] which leaves the impression that there must have been more ethnographic drawings than have survived in the Gordon Atlas.

[104] NA, Fagel-Archief, no. 2533, fol. 12, 24 April 1779.

[105] British Library, Add. MS 23.920, 24–26; a fourth drawing (no. 23) is a landscape. Forbes assumes that the drawings ended up in the British Library via Joseph Banks (cf. Forbes, 1951–2, pp. 87–89), with copies of the drawings; the copies are also included in Forbes (1965).

[106] British Library, Add. MS 23.920, 24.

CHAPTER SIX

THE ADVENTURES OF A SURINAMESE FRENCHMAN IN SOUTH AFRICA: THE TRAVEL ACCOUNTS OF FRANÇOIS LE VAILLANT

The opposite of Peter Kolb in the eighteenth-century literature on South Africa is the Frenchman François Le Vaillant (1753–1828), who was born in Surinam, in South America.[1] In contrast with Kolb, who presented an encyclopaedic and in several respects exhaustive collection of facts, Le Vaillant offers exciting travel stories—the *Voyage de M. Le Vaillant dans l'intérieur de l'Afrique* of 1790 and the *Second Voyage de M. Le Vaillant dans l'intérieur de l'Afrique* of 1795. The contrast between Kolb and Le Vaillant already catches the eye when the frontispieces of their books are put side by side. The purpose of a frontispiece is to present a striking illustration of the contents of the book.[2] The frontispieces of the books by Kolb and Le Vaillant show two ways of representing the African world.

The frontispiece of the Dutch translation of Kolb's work in particular represents the epistemology on which the book is based (fig. 26).[3] In the picture (designed by I.C. Philips) everything is characterised by objectivity, accurate representation and reliability. The allegorical action is performed mainly by three female figures. In the foreground is "Landkunde" (Geography) busy compiling a map, the same map that was included in the Dutch translation of Kolb's book. She is surrounded by putti with drawings—scaled-down illustrations from the book—and instruments that help to produce faithful representations of reality, such as a compass and a spyglass. Seated below the column bearing the title of the book is "Historikunde" (History), the descriptive science that is guided by the third female figure in the picture, "Ondervinding" (Experience). This figure has a touchstone in her hand, the symbol of the ability to determine the truth. Around her arm she is wearing a

[1] I use the most common spelling of Le Vaillant's name (he is also sometimes referred to as Levaillant).

[2] Donat (1966, pp. 169–170); Corbett & Lightbown (1979); Becker-Moelands (1985, p. 9); Bartelings (2001, pp. 245–247).

[3] Kolb (1727).

banderole with the barely legible Latin motto "rerum magistra", meaning that experience is the teacher in all things. Epistemologically, she is the central figure, because according to the "Verklaaring der titelprent" by Claas Bruin (1671–1732), which contains an explanation of the illustration on the frontispiece, History has "haare ogen sterk gevest/Op de *Ondervinding* die haar nadert... Zy, aller dingen leermeestres" (her eyes firmly fixed on *Experience*, who is approaching her... She, the teacher in all things). What History writes is given to her by the (personal) observation of reality, the 'autopsy' that Kolb emphasised so much. The aspects of reality described by History are depicted around the allegorical figures: Table Bay with Table Mountain in the background, a lion, some Khoikhoi. The user of Kolb's book must get the message that what he is about to read and see in the book is a faithful copy of the Cape world.

The frontispiece of Le Vaillant's *Voyage* (1790) is completely different (fig. 27).[4] Although part of the foreign world is depicted here as well in the form of a giraffe in an indistinct landscape with exotic fantasy trees, it is especially the gesticulating and conspicuously dressed figure of the author in the foreground that catches the eye.

The frontispiece in the *Voyage* refers to an important experience of Le Vaillant's, although it will only be described in his second book, the *Second Voyage* (1795): the shooting of a giraffe in the southern part of the present-day Namibia. Le Vaillant was immensely proud of this and describes at great length how he went about preserving parts of the giraffe.[5] Finding a giraffe was the main objective of his second journey to the north. He was hoping to make a famous contribution to natural history by shooting a giraffe.[6] There was a good chance that he would become famous, because giraffes (*Giraffa camelopardalis*) were still little known in Europe even after the publication of the *Nieuwste en beknopte beschrijving* in 1778. Descriptions, illustrations and parts of the body of a giraffe were sure to rouse substantial interest.[7]

---

[4] The caption to the title picture is: "Campement dans le Pays des Namaquois".

[5] Le Vaillant (1795, II, pp. 53–62).

[6] Le Vaillant (1795, II, pp. 1–2).

[7] In 1791 a German visitor called the skin of Le Vaillant's giraffe the main attraction of his collection of natural history artefacts. As the ceiling was too low, however, the skin could not be displayed on a display stand ("Nachricht von dem Naturalienkabinet des Hrn. le Vaillant", 1791, p. 95).

Visually more important than the giraffe is the extravagant Le Vaillant figure in the foreground, with exuberantly plumed hat, rifle, pistols in his belt and obedient hunting dogs. His pointing gesture is probably meant to guide the attention of the beholder. The Italian art theoretician Alberti (1404–1472) already recommended the use of pointing figures to draw the beholder's attention to a part of the representation.[8] The gesture draws attention to the giraffe, but can also be interpreted as an instruction to the dogs to attack the giraffe.[9]

Nevertheless it is especially the Le Vaillant figure that attracts the beholder's attention. He is in the foreground, is drawn more sharply and larger than the giraffe he is pointing at. The main difference between the Kolb and Le Vaillant frontispieces is that Le Vaillant does not appear to present his book as an objective copy, but as a personal experience. Whereas a portrait of the author as a man of learning is also included in the different editions of Kolb's *Capvt Bonae Spei Hodiernvm*, it is separated there from the representation of reality. On Le Vaillant's frontispiece, the author is integrated into the representation of reality; moreover, he is clearly in charge. The beholder's gaze follows the hand of the dominant Le Vaillant figure, just as it is his persona as narrator in the travel account that makes it possible for the reader to experience the foreign world as a series of *experiences* of the author. The giraffe in the frontispiece is a giraffe that is about to be shot down by the author as a unique contribution to natural history. It is Le Vaillant's giraffe, an episode from *his* adventures and a giraffe with which *he* is going to enrich natural history, the basis of *his* fame. This is also the way in which Le Vaillant's *Voyage* was introduced in the journal *Correspondance Littéraire* in 1789 shortly before its publication. The anonymous author of this review says that when one reads Le Vaillant's book "one participates in his [Le Vaillant's] amusement, his perils, his success; it is as if one is accompanying him on his journey, and one finishes the book as if one has really shared part of his life with him".[10] Because of the uncommon emphasis on subjectivity in the representation, Le Vaillant's

[8] Cf. Gandelmann (1986, p. 34).

[9] David Bunn, on the other hand, interprets the raised arm of the Le Vaillant figure as a gesture of the exercise of colonial power (Bunn, 1994, pp. 128–136). In this interpretation Bunn does not discount the obvious possibility that the gesture has an indexing meaning with regard to the beholder or that the Le Vaillant figure is giving a command to his dogs within the fiction of the representation.

[10] Cited in Contarini (2004, pp. 260–261).

book brought about a small French Revolution in the scholarly travel literature in terms of form.[11]

Popularity

Le Vaillant was not just an ordinary Frenchman. He was born in Paramaribo, in the Dutch colony of Suriname, in 1753 and lived there until he was ten years old, leaving with his parents for France in 1763. Although he spent only his childhood years in Suriname, he put great emphasis on his Surinamese descent in his books. He liked to present himself as an "American Creole".[12] Perhaps this makes him the first self-conscious Surinamese.[13] He had a reason for this too. Among other things, his descent had to justify his resistance to prevailing scholarly views and customary styles of writing, as I intend to show below. In this way Le Vaillant was an early postcolonial author in the sense of somebody who 'wrote back at' the establishment from a position that was, in his case, largely imaginary in a colony. He constructed an identity for himself as a Creole in order to enable himself to elude established European scientific and literary conventions in his narrative.

Between 1781 and 1785 Le Vaillant made two long journeys through South Africa, first to the east of the Cape Colony and then to the north, perhaps up into Namibia. The account of his travels appeared in two books, the *Voyage de M. Le Vaillant dans l'intérieur de l'Afrique, par le Cap de Bonne-Espérance, Dans les Années 1780, 81, 82, 83, 84 & 85* of 1790, which described his journey to the Eastern Cape, and the *Second Voyage de M. Le Vaillant dans l'intérieur de l'Afrique, par le Cap de Bonne-Espérance, dans les Années 1783, 84 et 85* in the fourth year of the revolutionary calendar (1795), which contains a description of his journey to the border region between South Africa and Namibia. Both books became bestsellers.

His Surinamese descent will have stood him in good stead in the Cape Colony during his journeys. Unlike some of his travelling contemporaries, such as the Swedes Sparrman and Thunberg, he seems to

[11] In the nineteenth-century travel literature the heroic explorer became a commonplace phenomenon (cf. Ryan 1996, pp. 21–53).

[12] Le Vaillant (1790, p. 202).

[13] When on 3 May 1772 Le Vaillant signed on as a cadet officer in the regiment of Berry, the entry noted that he was "natif de Surinam" (born in Surinam) (Le Vaillant, 2007, I, p. xiii), which suggests that his real-life identity was also Surinamese.

have had no problems making himself understood in the Cape Colony. But in part because of his Surinamese childhood his French was not good enough, and his manuscripts had to be edited for publication. His first account was probably edited by Casimir Varron (1761–1796), a Parisian man of letters, and his second by his French-born father. Since Le Vaillant's personal travel journals have not survived, the extent of the influence of these editors on the texts is not clear.[14] Besides these two travel accounts Le Vaillant also published a beautifully illustrated *Histoire naturelle des Oiseaux d'Afrique* in six volumes between 1796 and 1820 which earned him an important position in the history of ornithology—not only in South Africa, but also internationally.[15]

Le Vaillant's travel accounts were tremendously popular. Besides several editions in French, translations also appeared in German, English, Dutch, Italian, Swedish and Danish, sometimes in several editions.[16] And Le Vaillant himself was controversial. According to nineteenth-century ornithologists, he had fabricated various bird species in his *Histoire naturelle des Oiseaux* himself.[17] Subsequent travellers also doubted very much that he had penetrated as far to the east and north of southern Africa as he had claimed in his travel accounts.[18] Twentieth-century publications are more inclined to give him credit, however. The historian of geography Forbes cannot exclude the possibility that Le Vaillant had in fact visited areas far inland;[19] the historical zoologist Rookmaaker does his utmost to defend Le Vaillant's ornithological work.[20]

Le Vaillant's huge popularity until well into the nineteenth century indicates that his writings possessed the qualities necessary to reach a wide readership. One of the leaders of the German *Aufklärung*, the man of letters and geographer Georg Forster, gave high praise to Le Vaillant's *Voyage* of 1790. Forster wrote no less than two extensive reviews, in the *Göttingische Anzeigen* (1790) and the *Allgemeine Literatur-Zeitung* (1791), in which he held up Le Vaillant as the model for contemporary travel

---

[14] Le Vaillant (2007, pp. xliv–xlviii).

[15] Cf. Stresemann (1951, pp. 89–103).

[16] Ogilvie (1962) presents a list of the editions and translations of the *Voyage* and the *Second Voyage*. In this chapter I refer to the first quarto edition of the *Voyage* and *Second Voyage*. Of the *Voyage* there is also a slightly enlarged later edition—especially with more pictures. However, I used the 1790 edition because this was the version on which the translations were based.

[17] Rookmaaker (1989, pp. 188–189).

[18] Forbes (1973).

[19] Forbes (1973, p. 89).

[20] Rookmaaker (1989, pp. 189–190).

literature.[21] Forster's opinion carried a great deal of weight in north-western Europe because he was professionally well informed on travel literature and had made an important contribution to the genre himself with his account of a voyage to the Pacific in the company of James Cook (*A voyage around the world* (1777)) and his *Ansichten vom Niederrhein* (1790), a book regarded as groundbreaking in the history of German literature.[22] Moreover, Forster also knew Cape Town from short visits in 1772 and 1775 in the course of his voyage around the world with James Cook.

Forster found especially the format of Le Vaillant's *Voyage* exceptional. About the content he did not say much. There was not much in the *Voyage* that was not already known from other sources, even though what Le Vaillant had to offer in this regard was worth while in its own right.[23] What struck Forster in particular, however, was the mould in which it was cast, so much so that he even adapted his own views of travel literature to those of Le Vaillant afterwards.[24] According to Forster, Le Vaillant succeeded in captivating his reader right up to the end by telling a story instead of inundating him with an incoherent amount of information, as happened all too often in other travel accounts:

> Once one has opened this travel account, one cannot put it away unread, and during the reading the thread of the narrative runs so smoothly, without any hiccoughs, that the reader remains conscious of a sense of an indivisible, spirited whole, as it were, from start to finish. The works of tastless industry, however, show no sign of this creative imprint but are lifeless compositions [...][25]

According to Forster, Le Vaillant offers a synthetic, "spirited" and well-composed total picture. With amusing anecdotes about his domestic

[21] Both reviews are included in *Georg Forsters Werke* (Forster, 1992, XI, pp. 225–230, 270–278). Georg Forster's father, Johann Reinholt Forster, had translated Le Vaillant into German.

[22] Hentschel calls Forster the most important eighteenth-century expert on travel literature, at any rate in Germany (Hentschel, 1999, p. 46).

[23] Forster (1992, XI, pp. 226, 275).

[24] Hentschel (1999, p. 56).

[25] "Wenn man diese Reisebeschreibung einmal in die Hand genommen hat, kann man sie nicht ungelesen wieder weglegen, und indem man liest, läuft der Faden der Erzählung so ohne allen Anstoß, ohne alle Unebenheiten fort, daß die Vorstellung eines Unzertrennlichen, gleichsam beseelten, Ganzen dem Leser von Anfang bis zu Ende gegenwärtig bleibt, dahingegen die Werke des geschmacklosen Fleißes von diesem schöpferischen Gepräge nicht die geringste Spur verrathen, sondern todte Zusammensetzungen sind [...]" (Forster, 1992, XI, p. 225).

pets and especially about the Gonaqua girl Narina, whom Le Vaillant adored, he keeps up the suspension (p. 228). Even if Le Vaillant's description of his affair with Narina were prettier than it actually was, Forster has no objection (p. 228). In his second review Forster puts it even more strongly. Here he calls Le Vaillant's travel account "a work of art" and the episode with Narina a "small novel" (p. 275). He now puts all the emphasis on the literary significance of Le Vaillant's description.

In the discussion below I want to scrutinise mainly two facets of Le Vaillant's travel account. First, Le Vaillant's literary presentation, which Forster likes so much: Le Vaillant's ability to turn the very diverse experiences and observations into a personal experience that enables the reader to experience southern Africa together with him. In this respect his work had a significance that rose above his contribution to the representation of South Africa. His books changed the conventions of European travel literature.

In the second place I want to examine his scholarly aspirations. Le Vaillant was certainly esteemed by some contemporaries for the knowledge he had collected. With certain reservations, the nineteenth-century travellers John Campbell (1766–1840) and William Burchell (1782–1863) were prepared to recognise the authority of Le Vaillant as an ethnographer. Even if according to Campbell he was "romantic", i.e. extravagant or imaginative, his descriptions of the Khoikhoi were nevertheless extremely valuable: "yet he gives the best account of the manners and customs of the Hottentots I have seen".[26] Indeed, despite all his jokes and antics Le Vaillant claimed scientific authority, and his contributions were esteemed as such by contemporaries.[27]

---

[26] Campbell (1974, p. 330).

[27] The older literature occupied itself mainly with the factual truth of Vaillant's work (Grant, 1957; Forbes, 1965, 1973). The more recent literature also deals with ideological issues, but like Pratt (1992, pp. 88–90) it attends to Le Vaillant only in passing or gives inaccurate indications, like Boisacq (1993), who makes vague connections between Le Vaillant on the one hand and Diderot and Rousseau on the other. Ian Glenn, however, published a valuable article on the editorial problems surrounding the *Voyages* [...] (Glenn, 1997). Contarini (2004) wrote an article on Le Vaillant's contributions to natural history and anthropology. As regards the latter, she incorrectly relates his ethnographic descriptions to Degérando's *Considérations sur les diverses méthodes a suivre dans l'observation des peuples sauvages* (1800), which was published after Le Vaillant's voyage and travel accounts and is far removed from Le Vaillant's ethnographic practice in other respects as well (see Chapter 9 for Degérando). Le Vaillant's contribution to natural history in the eighteenth century is examined in Rookmaaker et al. (2004). Popular publications, such as *The truth in masquerade, the adventures of François le Vaillant*

Aesthetisation

In the main, Le Vaillant's travel accounts follow the chronology of events—the "natural presentation" of a travel account, as the Swedish traveller Anders Sparrman calls it.[28] Here and there the text of journal entries can still be discerned, for example in the dates, which are mentioned fairly randomly. However, for publication the whole was edited into a continuous travel narrative in order to optimise its entertainment value.

Le Vaillant is very conspicuously present in his book as narrator, focaliser and person. One literally cannot overlook him, as he is also shown in many illustrations in the book.[29] In different pictures he is seen moving about in a plumed, wide-brimmed hat; where he is not present in person, one sees his attributes: his tent and his wagon, his domestic animals, his servants. Almost all the topographic representations therefore suggest that Le Vaillant was present on site, in front of the Heerenlogement, in the Olifants River, in Namaqualand, on the banks of the Orange River.

The information content of pictures of this kind and the accompanying stories is often slight. The contours of the landscapes in the pictures are usually barely recognisable, if at all. The historian of geography Forbes had trouble identifying the equivalent of some of Le Vaillant's topographic drawings.[30] I was personally able to determine that the correspondence between the landscape at the Heerenlogement—where the situation has changed little since the eighteenth century—and Le Vaillant's representation of it in his *Second Voyage* (Pl. iv) is minimal. It therefore seems as if the pictures have an illustrative rather than an

(1973) by Jane Meiring do not go much further than recounting the anecdotes told by Le Vaillant himself. In 2007 the Van Riebeeck Society published the first volume of a new English edition of the *Voyage* translated by Ian Glenn and Ian Farlam with a valuable introduction (Le Vaillant 2007).

[28] Sparrman (1974–1977, I, p. 36).

[29] The authorship of these illustrations is even less clear than that of the text of the travel account. In any case they hark back to a more extensive series of coloured drawings kept in the Library of Parliament in Cape Town, which according to the art historian Bokhorst (1973, pp. 103–127) was produced through the collaboration of various draughtsmen supervised by Le Vaillant. In his text Le Vaillant claims that they are based on his own sketches. However, these sketches have not survived; neither has the diary Le Vaillant kept during his voyage.

[30] Forbes (1973).

informative value.[31] They depict the experiences of Le Vaillant in a certain place—illustrate them—more than they reflect the landscape. The story and illustration of Le Vaillant's crossing of the Olifants River in the Western Cape will serve as an example. In the *Second Voyage* Le Vaillant recounts with self-deprecation how he crossed this river in the western part of South Africa. As he was unable to swim, it became a "grotesque" performance. Sitting astride a tree trunk as if it were a saddled horse, he had himself pulled across by his indigenous servants (called "plaisans Tritons" in the travel account). The episode is captured in a drawing (fig. 28). We see Le Vaillant in the foreground, wearing his plumed hat and in full attire, sitting on his floating tree trunk in the raging torrent, pulled along by his servants. It does not leave much room for the surrounding landscape, which is nothing more than a background to the floating traveller.[32] The story and the picture seem to be intended more to entertain the reader than to reflect a part of South Africa.

Where the landscape receives more attention, as in the picture of "Camp de Pampoen-Kraal" (fig. 29),[33] it is all about only one aspect of what is little more than a clichéd landscape, a natural "cavern" in the bush that was used as a shelter by Le Vaillant and which he praised in the text as being superior to the artificial garden architecture in Europe.[34] The foreground and even the background are densely populated with miniature replicas of a Khoikhoi figure shown elsewhere in the book. Le Vaillant's tent and wagons are also shown, with his pet baboon Kees on the hood of the wagon and he himself, with plumed hat and rifle, on a miniature scale exactly in the centre of the drawing. It seems as if the entire exotic world is arranged around Le Vaillant, the pivotal figure. The exotic world is the backdrop for the adventures of Le Vaillant the traveller.

Le Vaillant employed two rhetorical strategies that ensured that his traveller's tales could fascinate even critical contemporary readers such as Georg Forster. In the first place, as we saw, he constructed a central figure in his narrative and in illustrations with whom readers

[31] I follow a distinction here that was made by Bernard Smith (Smith, 1992, p. 53).

[32] *Second Voyage*, I, pp. 124–8; "Passage de la Rivière des Éléphants" (Crossing of the Olifants River), Pl. v.

[33] Le Vaillant (1790, opposite p. 85).

[34] Le Vaillant (1790, p. 85).

could identify. His traveller persona is not reduced to a "seeing man", a traveller who has little substance as a person and mainly records the environment;[35] instead he is a person with identifiable attributes and feelings who can experience love, loneliness and melancholy—emotions that would have upset the mind of the scientific researcher but helped to offer the sensitive eighteenth-century readers who were familiar with Goethe and Rousseau a person they could identify with.[36] The great advantage of this strategy is that it prevents the fragmentation that made so many eighteenth-century travel accounts so difficult to read. The perspective of the personal narrative keeps the excessive variation of events and things worth knowing in check. In the second place he offers entertaining and not entirely credible tales ranging from the farcical to the insipidly sentimental. As Forster already pointed out, stories of this type guaranteed suspense. In many cases these two strategies overlap, partly because many of the entertaining anecdotes concern Le Vaillant and his personal environment. The anecdotes concern for example his beard and clothing, his tame baboon Kees, his cockerel (which he had taken along as a portable alarm clock), his faithful servant Klaas and his love for the Gonaqua girl Narina. In addition he serves up hunting adventures, stories of hardship suffered and threats from bandits—the classical ingredients of adventure novels.

The most sensational tale is undoubtedly the story of his love for Narina. Forster already found this beautiful, but doubted its truth. It is indeed doubtful whether this story is entirely true. The always very serious English traveller John Barrow—"Mr Chronometer"—already had his doubts and even launched an investigation. When he visited the area where the Gonaquas had lived, nobody was able to tell him anything about Narina. Even her name was an unknown word in this region.[37] In any case, the entire Narina episode is of such a bucolic character that this tale does seem to have sprung from Le Vaillant's literary imagination.

Le Vaillant said that he noticed Narina for the first time during the exchange of gifts with the chief of the Gonaquas. At the time she was a girl of sixteen who acted rather reticent at the first meeting. Le Vaillant

[35] Le Vaillant (1790, p. 85).

[36] Le Vaillant once addresses his reader as "Le Lecteur sensible" (the sensitive reader) (Le Vaillant, 1795, I, pp. 204–205). Also compare Rookmaker et al. (2004, pp. 101–103).

[37] Barrow (1801, p. 227).

found her attractive. She had the most beautiful teeth in the world. With her willowy figure she would have been a suitable model for the painter Francesco Albani.[38] "She was the youngest of the three Graces in the embodiment of a Khoikhoi woman". "The domain of beauty" seemed to extend to the Khoikhoi.[39] Narina satisfied the codes of the classicist ideal of beauty. The accompanying picture had to confirm this. In the picture, which purports to be a portrait of Narina (fig. 30), she is presented in *contrapposto*: the weight on one leg, the other leg slightly bent and with a bulging hip, a pose associated in painting and sculpture with grace and people of the higher classes.[40] Her pose also suspiciously resembles that of the Venus de' Medici, the yardstick of female beauty for many people of the eighteenth century.[41]

As Le Vaillant tells it, Narina was somewhat coy when they met for the first time, but after he had given her some gifts she became more approachable. However, the affair remained limited to mutual teasing and flattery. He teases her when she goes swimming with her friends by sitting down on their clothes on the bank. She responds with "playful nudges". He shoots birds for her instead of for his collection,[42] but he nevertheless maintains his "modesty" towards her and sends her to her mother when she wants to go to sleep.[43] For readers accustomed to eighteenth-century novels, the description of these pastoral scenes made titillating reading. Some readers may also have been aware of the ambiguity of birds that were shot.[44]

The Narina story serves a double purpose. In the first place it livens up the travel account. Le Vaillant uses the story as the framework of a long description of the manners and customs of the Gonaquas and the shorter description of the Xhosas. Once these instructive but somewhat dry subjects have been disposed of, the reader is rewarded with a short farewell scene between Le Vaillant and Narina. However, the story also

[38] The reference is to the painter Francesco Albani of Bologna (1578–1660), a painter of idyllic landscapes peopled by gracious figures. Albani (called Albane in French) was much admired in the eighteenth century (Puglisi, 1996).

[39] Le Vaillant (1790, p. 186).

[40] Cf. Roodenburg (sa, pp. 120–127).

[41] Haskell & Penny (1982, pp. 325–328); Bindman (2002, p. 126).

[42] Le Vaillant (1790, pp. 191–193).

[43] Le Vaillant (1790, p. 189).

[44] Referring to 'vogelen' ('birding'), which can mean in Dutch both hunting birds, catching birds with nets or in a cage and sleeping with someone (*Woordenboek der Nederlandsche Taal op CD-Rom*). The open bird cage in genre paintings (ICONCLASS 33C4212) is symbolic of the loss of virginity.

has a function in the debate about the status of the Khoikhoi. Kolb had already done much to raise their moral standing, but as regards aesthetics he shared the prevailing view that Khoikoi women were the ugliest on earth.[45] The Narina story and the accompanying drawings prove the contrary. Khoikoi women can be pretty, and there are attractive male specimens too. They also receive the classicist treatment to make them acceptable for the European public (fig. 31). Le Vaillant calls the men Jugurtha and Syphax, after Nubian heroes from Livius's description of the Punic Wars.[46] Le Vaillant made the Khoikoi, in the form of the Gonaquas, morally superior and physically attractive to boot. Thus he completed the rehabilitation of the Khoikoi which Kolb had already advanced so far.

Moral hierarchy

The *Voyage* and the *Second Voyage* are not only egocentric adventure stories. The books also contain "excursions" embedded in the travel account,[47] varying in length and accompanied by illustrations of the botany, zoology and especially the ethnography of southern Africa, in which Le Vaillant enters into discussion with the existing literature. Le Vaillant's books also have scientific pretensions. Within the descriptive tradition of eighteenth-century science, they want to present facts about animals as yet unknown, such as the giraffe, and Khoikhoi in their unspoilt state. Like Kolb, Le Vaillant sees himself as a "historian" in this field, which in the Plinian sense of the *Historia naturalis* means someone who records facts: "I wish to reiterate that I am only a historian and that I do not wish to prove anything or to explain facts".[48] Below I will confine myself to his treatment of ethnographic facts.

---

[45] Even Kolb expressed his aversion to the appearance of Khoikhoi women, especially if they had a Hottentot apron, whereas he describes the men's faces as friendly (Kolb, 1719, pp. 425, 371). During the second half of the eighteenth century the image of the Khoikhoi among the French Philosophes remained ambiguous. Besides the innocent Hottentot who owed his existence to Kolb's book, the monstrous Hottentot persisted (Fauvelle-Aymar, 2002, pp. 285–303).

[46] "Amiroo", Narina's cousin (Le Vaillant, 1790, p. 197) and "Jugurtha" and "Syphax" (Le Vaillant, 1795, II, p. 156).

[47] Le Vaillant (1795, I, p. 44).

[48] Le Vaillant (1795, I, p. 42); Seifert (1976).

Le Vaillant is primarily interested in 'savage' peoples living far from the colonial settlements. This is why his ethnographic descriptions in the *Voyage* commence only once he has reached the Gonaquas on the eastern border of the Cape Colony. The description he gives of the Gonaquas is the most extensive ethnographic description of his two travel accounts and serves as a foundation for the subsequent shorter descriptions of other peoples, in the sense that in later descriptions the indication of similarities with and differences from the Gonaquas will often suffice.

As it did for Gordon, the existing literature—and Kolb in particular—serves as a guideline for the descriptions and supplies a checklist of subjects to be dealt with.[49] At the same time Le Vaillant takes issue with this literature. For example, at the first opportunity when Kolb comes up, he recounts an embellished version of La Caille's slanderous story of Kolb's laziness.[50] It was said at the Cape, he says, that towards the end of his stay at the Cape Kolb went from pub to pub to have drunkards tell him the absurd tales that ultimately found their way into his book.[51] Nevertheless we again find the classic subjects from the discourse on the Cape in Le Vaillant, such as the description of the 'Hottentot apron'—this time illustrated with a portrait of a woman sporting this feature (fig. 3)[52]—and the partial castration of Khoikhoi men, where he is not in the position to refute Kolb's version completely.[53] Even the story of Narina's beauty is inspired by the existing literature, because it constitutes a repudiation of the aversion to the appearance of Khoi women that was expressed by all preceding authors.

What is new, however, is that Le Vaillant uses a primitivist framework to weigh up the different population groups at the Cape against one another. He declared some groups, such as the Gonaqua, to be noble savages,[54] possessing admirable moral qualities that the Europeans had lost. Broadly speaking, this framework leads to appreciative remarks

[49] Sometimes he also states this explicitly: "Je leur faisoit sur-tout beaucoup de questions sur Kolbe & différens Auteurs; sur leurs religions, leurs loix, leurs usages. Ils me rioient franchement au nez" (In particular I put many questions to them about Kolb and various authors; about their religion, their laws, their customs. They simply laughed in my face.) (Le Vaillant, 1790, p. 70).

[50] Cf. chapter 2.

[51] Le Vaillant (1790, p. 48).

[52] Le Vaillant (1790, p. 370 with picture on Pl. viii).

[53] Le Vaillant (1795, II, pp. 290–293).

[54] Le Vaillant (1790, pp. 253–254, 257; 1795, II, p. 183).

about indigenous people and criticism of the colonial dispensation and the individuals it has produced.

Primitivism—the glorification of natural man as an example of human perfection[55]—had already become a commonplace by the end of the eighteenth century and was reflected in a range of cultural products. In eighteenth-century Europe, the noble savage appeared on the stage and even sang in the opera.[56] However, the concept had acquired a particular significance after 1755 in Rousseau's representation of the primitive existence in his *Discours sur l'origine et les fondemens de l'inégalité parmi les hommes* (*Discours sur l'inégalité* hereafter).[57] The subject of this text is the contrast between the goodness of natural man and the moral perversity of the egoistic civilised person. In his *Discours sur l'inégalité* Rousseau describes the decay of human happiness in the course of the history of mankind. To summarise, he says that after a period of inane but satisfied animal existence, people came together to form primitive hunter societies and found a happy equilibrium between nature and culture, which was disturbed as soon as man started tilling the soil and working metal. From that moment onward, inequality between people increased and civilised man came into being, a monster of corruption and malevolence. For empirical data Rousseau had made use of, among others, Kolb's description of the Khoikhoi.[58] On the frontispiece of the first edition of Rousseau's *Discours sur l'inégalité* the moment is shown where this Khoikhoi, in front of Commander Simon van der Stel and his compatriots, returns his European clothes in exchange for a quasi-

[55] In my use of the term primitivism I follow the broad definition given by Christian Marouby: "le primitivisme [...] c'est [...] la découverte d'une harmonie originelle, d'un état qui, loin de menacer l'homme d'une chute dans l'animalité, semble le rapprocher de sa perfection, de son humanité même" (primitivism [...] is the discovery of an original harmony, of a state that, far from threatening man with falling into animality, seems to bring it closer to perfection, even to his humanity) (Marouby, 1990, p. 122).

[56] Cro (1990, pp. 131–134).

[57] Below I refer to the modern standard edition of the *Discours*: Jean-Jacques Rousseau, *Oeuvres completes* (1964, III, pp. 110–240). I will not go into the complex relation between Rousseau and primitivism. Views of this relation in the literature about Rousseau depend greatly on the definition given to the concept of primitivism. Nevertheless, Rousseau's picture of an ideal existence—mankind's 'golden age' that was said still to exist among primitive peoples—is most certainly primitivistic if the definition of primitivism given by Marouby (1990) quoted above is followed. For a different opinion I refer to the article by Arthur O. Lovejoy, "The Supposed Primitivism of Rousseau's Discourse on Inequality" (Lovejoy, 1960, pp. 14–37), which still has considerable influence.

[58] Pire (1956, pp. 355–378).

indigenous loincloth.[59] Kolb's honourable Khoikhoi, further upgraded to a noble savage by Rousseau, returns to South Africa in this new embodiment in Le Vaillant's *Voyage*.

Although Le Vaillant does not mention Rousseau, there are distinct echoes of the *Discours sur l'inégalité* in Le Vaillant's books.[60] Obviously, first of all there is the thought that man in his natural state is morally superior to the eighteenth-century bourgeois, although this thought is also found among other supporters of primitivism.[61] More specifically dependent on Rousseau is Le Vaillant's theory that primitive peoples still offer some impression of the human species in its "childhood".[62] Rousseau expressed this thought somewhat more subtly in his *Discours sur l'inégalité*, where he draws a distinction between the innocent animal existence of natural man living in isolation, who represents the "childhood" of mankind, and a happy mean between nature and culture that was found when people came together into primitive communities, which can be regarded as the "youth" of mankind. The crucial passage in the *Discours sur l'inégalité* about this "youth phase" is the following:

> Thus, though men had become less patient, and their natural compassion had already suffered some diminution, this period of expansion of the human faculties, keeping a just mean between the indolence of the primitive state and the petulant activity of our egoism, must have been the happiest and most stable of epochs. The more we reflect on it, the more we shall find that this state was the least subject to revolutions, and altogether the very best man could experience; so that he can have departed from it only through some fatal accident, which, for the public good, should never have happened. The example of savages, most of whom have been found in this state, seems to prove that men were meant to remain in it, that it is the real youth of the world, and that all subsequent advances have been apparently so many steps towards the perfection of the individual, but in reality towards the decrepitude of the species.[63]

---

[59] See fig. 11.

[60] Boisacq (1993, pp. 117–131) also pointed out in a different way parallels between Le Vaillant and Rousseau. Le Vaillant even named one of his sons after Rousseau in 1793.

[61] Le Vaillant (1790, pp. 253, 257).

[62] Le Vaillant (1790, p. 255).

[63] Translation: G.D.H. Cole: http://www.constitution.org/jjr/ineq.txt. "Ainsi quoique les hommes fussent devenus moins endurants, et que la pitié naturelle eût déjà souffert quelque altération, cette période du développement des facultés humaines, tenant un juste milieu entre l'indolence de l'état primitif et la pétulante activité de notre amour-propre, dut être l'époque la plus heureuse et la plus durable. Plus on y réfléchit, plus

With his enthusiastic description of the happy existence of the Gonaquas Le Vaillant wanted to demonstrate the same as Rousseau—not in the form of speculative historiography, but as observed reality. He supplied quasi-empirical evidence for Rousseau's speculation. This operation had less to do with creating respect for the Gonaquas' own culture and nature than with European speculation about the best form of society. In order to render the Gonaquas serviceable for this purpose, Le Vaillant presented them in a classicist fashion, as we saw above.

One can find further borrowings from Rousseau in Le Vaillant. He observes that the primitive hunter communities are egalitarian. In war there are no ranks; everyone is general and soldier at the same time.[64] Furthermore, according to the *Discours sur l'inégalité* all primitive communities are egalitarian until agriculture and metal working appear on the scene. Also, according to Le Vaillant, primitive man experienced no jealousy in matters of love,[65] a thought that is also found in Rousseau, although the latter assigns the absence of jealousy more to the most primitive period when people did not yet live together in communities.[66] An important similarity is Le Vaillant's negative judgment of colonial society at the Cape, which is related to Rousseau's rejection of the "état civile". Cape colonial society is the local variant of civil society, which Rousseau detested. Cape society had, according to Le Vaillant, been created by deceit. He asserted that the founder of the Cape Colony, Jan van Riebeeck, had employed all possible subterfuges to gain the goodwill of the Khoikhoi to wheedle them out of their land. "He had coated the brim of the poisoned chalice with honey".[67] Other charges against Cape colonial society were at least in the vein of Rousseau. Deceit is the main characteristic of the colonists in the immediate surroundings of Cape Town, who are obsequious and haughty at the same time.[68] The trekboere (itinerant farmers) are the guilty party in

on trouve que cet état était le moins sujet aux révolutions, le meilleur à l'homme, et qu'il n'en a dû sortir que par quelque funeste hasard qui pour l'utilité commune eût dû ne jamais arriver. L'exemple des sauvages qu'on a presque tous trouvés à ce point semble confirmer que le genre humain était fait pour y rester toujours, que cet état est la véritable jeunesse du monde, et que tous les progrès ultérieurs ont été en apparence autant de pas vers la perfection de l'individu, et en effet vers la décrépitude de l'espèce" (Rousseau, 1964, III, p. 171).

[64] Le Vaillant (1790, p. 223).

[65] Le Vaillant (1790, pp. 226, 253).

[66] Rousseau (1964, III, p. 169).

[67] Le Vaillant (1790, pp. 129–130).

[68] Le Vaillant (1795, I, p. 24).

the border war with the Xhosas. In this conflict they have "committed all the atrocities thought up in hell".[69]

The children of European fathers and Khoikoi mothers, the "Bastaards", also called Bastard-Whites by Le Vaillant, are a particular product of colonial expansion in South Africa. This group possesses especially the characteristics of their fathers. They are more courageous, more energetic and more entrepreneurial than the Khoikhoi, but also more malicious. The mixing of Khoikhoi with "Negroes" had produced better results, individuals who are hard-working and faithful. The rare product of the relationship between a white woman and a Khoikhoi is also better, because the child then inherits the good nature of his father; in Le Vaillant's theory of heredity the father is the dominant party. Unfortunately, the last-mentioned two types are in the minority by far compared with Bastaards of the first type, because the mating of Khoikhoi women with whites produces a larger progeny, as white men are more active than Khoikhoi men. Le Vaillant therefore foresees that the "restless" race of the Bastard-Whites will ultimately govern the Cape.[70]

For the surviving primitive peoples the avarice of colonial society is a serious threat. Le Vaillant expresses it as an internal dialogue, which as it happens appears to echo a passage from Raynal's *Histoire des deux Indes*:[71]

> Happy mortals, I said to myself, retain this invaluable innocence for a long time still, but live unknown! Poor Savages, do not mourn having been born under a burning Sky, on arid and poor soil that barely produces heath and brambles. Alas, rather view your circumstances as a favour to which Heaven draws your attention. Your deserts will never arouse the greed of the Whites. Unite with the happy bands of people who are so lucky as not to know them [the Whites] better than you do. Destroy, eradicate to the minutest traces this yellow powder [gold] that develops into metal in your ravines and your rocks. You will be lost if they discover

[69] Le Vaillant (1790, pp. 146, 154–155).

[70] Le Vaillant (1790, pp. 262–266).

[71] Raynals *apostrophe* is aimed at the Hottentots: "Fuyez, malheureux Hottentots, fuyez! enfoncez-vous dans les fôrets. Les bêtes féroces qui les habitent sont moins redoutables que les monstres sous l'empire desquels vous allez tomber. Le tigre vous déchirera peut-être; mais il ne ôtera que la vie. L'autre vous ravira l'innocence & la liberté" (Flee, unfortunate Hottentots, flee! Hide yourselves in the forests. The ferocious beasts that inhabit them are less fearsome than the monsters under whose dominance you will fall. The tiger may perhaps devour you, but he will take only your life. The other will rob you of you innocence and your freedom) (Raynal, 1780, p. 399).

> it. Know that it is the scourge of the Earth, the source of all crimes, and especially prevent the approach of a De Almagro, a Pizarro, a Hernán Cortés and the blood-stained star of De Valdivia.[72]

However, Le Vaillant is not consistent in his application of the primitivistic scheme. In his *Second Voyage* in particular he is much less enthusiastic about primitive peoples. There he even uses an orthodox hierarchy of civilisation. Following his visit to the Namaquas in the south of Namibia, he asserts that the Khoikhoi are the least intellectually gifted people on earth.[73] Another people of this region has, by digging wells, risen above "the inferiority of the existence of the savage", and because of their nomadic way of life the Kabobiquois are superior to their neighbours, among others because they have a "more highly developed morality".[74] In the course of his second journey he discovers a hierarchy of civilisation in terms of which he ranks more highly those who more closely approximate the European norm. But colonialistic pure and simple is the maxim whereby he tries to assert himself among the savages: "The best policy to overawe the savages is to impress them with something exceptional that must convince them that the white race is superior to theirs".[75] Sometimes he even completely forgets his primitivistic scheme. This happens when he unfolds plans for the foundation of a new city with extensive fields and a port in the area of Vierentwintig Rivieren north of Cape Town, at Saldanha Bay, where he wanted to establish an agricultural colony. Typical colonial crops such as sugar cane, cotton and indigo would be cultivated on these fertile fields.[76]

[72] "Mortels heureux, me disois-je, conservez long temps cette précieuse innocence; mais vivez ignorés! Pauvres Sauvages, ne regrettez point d'être nés sous un Ciel brûlant, sur un sol aride & desséché qui produit à peine des bruyères & des ronces; regardez, ah! plutôt regardez votre situation comme une faveur signalée du Ciel; vos déserts ne tenteront jamais la cupidité des Blancs; unissez-vous aux peuplades fortunées qui n'ont pas plus que vous le bonheur de les connoître; détruisex, effacez jusque'aux moindres traces de cette poudre jaune qui se métallise dans vos ravines & dans vos roches; vous êtes perdus, s'ils la découvrent; apprenez qu'elle est le fléau de la terre, la source de tous les crimes, & redoutez sur-tout l'approche d'un Almagro, d'un Pizarre, d'un Fernand-Cortez, & l'étole ensanglantée des Vanverdes" (Le Vaillant, 1790, p. 257). The four conquistadores mentioned here by Le Vaillant are: Diego de Almagro (1518–1548), Francisco Pizarro (1480–1541), Hernán Cortés (1485–1547) and presumably Pedro de Valdivia (1500–1554).

[73] Le Vaillant (1795, II, p. 98).

[74] Le Vaillant (1795, II, pp. 165–166).

[75] Le Vaillant (1795, I, p. 235).

[76] Le Vaillant (1795, I, pp. 100–102).

Le Vaillant does seem to be aware of his inconsistency, however. When he leaves the Gonaqua in the course of his first journey, he first says that he is leaving with a heavy heart, because he will now again—in the spirit of Rousseau—have to allow himself to be fettered by the chains of civil society (the "Société").[77] This feeling of depression is reinforced when on his way back he finds the first signs of civilisation in the hovel of "Two Negroes" who receive him cordially. His heart is heavy. He passively accepts their kind assistance because he recognises the habits of civil society in them. At that moment he even starts missing the wild animals.[78] But a little later he admits that this was all pretence:

> All these vague and romantic ideas that had misled me, all the feelings of sadness with which I had fed my heart when I left the Savages, finally began to fade away and Reason regained the upper hand, which made it clear enough to me that I had to fulfil other obligations, as I had not been born for this nomadic and dangerous life.[79]

Among the obligations that forced him to return he was thinking in the first place of his close friendship with a colonial official, the fiscal Boers. This friendship and others with wealthy patrons from the colonial elite, which he mentions with pride, show up Le Vaillant's ambiguous position as a champion of natural man even more clearly. Le Vaillant says that his journey from the Netherlands had been made possible by Jacob Temminck (1748–1822), the Treasurer to the VOC in Amsterdam. Archival evidence reveals that Le Vaillant was registered as a "constabelmaat" (gunner's mate) in the service of the VOC during the time he was away from the Netherlands. After his return he was paid accordingly.[80] He remains silent about this detail in his two books though, where he only mentions that his close friend Temminck used

[77] Le Vaillant (1790, pp. 346–347). One is reminded of Rousseau's famous dictum in *Du contract social* about the condition of man in the "état civil": "L'homme est né libre, et par-tout il est dans les fers" (man is born free, and finds himself in chains everywhere) (Rousseau, 1964, III, p. 351).

[78] Le Vaillant (1790, pp. 380–381).

[79] "Tout les idées chimériques & romanesques qui m'avoient bercé, tous ces déplaisirs dont je nourrissois mon coeur en quittant les Sauvages, commençoient enfin à se ralentir, & la Raison reprenant le dessus, me faissoit assez connoître que, n'étant point né pour cette vie errante & précaire, j'avois d'autres obligations à chérir" (Le Vaillant, 1790, p. 392).

[80] Le Vaillant (2007, I, pp. xv–xvi). After he returned to Amsterdam Le Vaillant received 552 guilders, 2 stuivers and 4 penningen for his services as a VOC employee (Le Vaillant, 2007, I, p. xv, nt. 13).

his influence in order for Le Vaillant to obtain permission to travel in one of the Company's ships. All this might have been part of a secret agreement between Temminck and Le Vaillant. In exchange for patronage at the expense of the Company, Le Vaillant undertook to collect new specimens for Temminck.[81]

Le Vaillant also obtained support locally. After he had lost his possessions during a sea battle in the Fourth Anglo-Dutch War he received a completely new outfit from the Cape fiscal Willem Cornelis Boers (1745–1803).[82] He also received much assistance during the preparation of these travels from Robert Gordon, the commander of the VOC garrison at the Cape, in whose house he stayed for some time after the departure of Boers. On his journeys he used a letter of recommendation from the governor that instructed colonists to give assistance to Le Vaillant. As with Temminck, his friendships with colonial office bearers were based especially on a shared interest in the exotic nature of southern Africa. In exchange for the assistance he received from them Le Vaillant contributed to their collections of curiosities.[83] He dedicated the 1790 *Voyage* to Boers.

A natural naturalist

Despite his material dependence on colonial office-bearers, Le Vaillant creates for himself an exclusive position in his stories. This fits his persona as the hero of the story, which sometimes assumes the features of a picaresque hero extricating himself from all sticky situations. He projects himself as the traveller who can always inspire trust among the indigenous tribes, who wins a war, puts a stop to faction fights (for which they offer to make him chief of the tribe), reconciles family members who are at loggerheads and, of course, shoots animals *en masse.*

On the scientific side he claims an exclusive position outside and even in opposition to established scientific practice. The attention drawn to the traveller and his experiences has not only a technical narrative function. Le Vaillant also uses it to distance himself from his learned colleagues. He presents himself as an ordinary bloke without

[81] Temminck purchased the major part of Le Vaillant's African collections (Le Vaillant, 2007, I, p. 10, nt. 10).

[82] Le Vaillant (1790, p. 382). For Boers see Naudé (1950).

[83] Rookmaaker (1989, pp. 256–271).

the pretensions of (in his eyes) armchair scholars such as Kolb. The reader must understand that Le Vaillant is someone who would rather tell amusing anecdotes about the misbehaviour of his tame baboon than bore the reader with learned zoological descriptions.[84] This tenor is present in his account right from the start. The report of his first journey (in the *Voyage*) is preceded by a description of the evolution of Le Vaillant, the young researcher of nature. Travellers almost always include an account of their journey out and back, as Le Vaillant also does, but according to Le Vaillant it is somewhat unusual to start a travel account with the traveller's birth.[85]

> It may perhaps be found strange that, in order to relate the account of a journey recently undertaken in Africa, I was forced to revisit the past and take my Readers to South America to trace the first steps of my childhood. However, it did not strike me as injudicious to justify, by recounting the early days of my life, my way of seeing, thinking and acting, which will always retain the flavour of my land of birth and which, if perhaps judged severely, will not fail to shock those intolerant minds that never take kindly to being stripped of their prejudices and to someone daring to launch a frontal attack on those principles and practices which are generally accepted.[86]

His Surinamese descent must explain why his conduct and views deviate from the norm. He therefore sees fit to bring up this matter at the start of his travel account. However, Le Vaillant does not confine himself to merely mentioning his country of birth; in a "Précis Historique" in the *Voyage* he also briefly describes his scientific development from his birth in Dutch Guyana (Surinam), a region where exceptions to the rule of nature are found, which perhaps also implies that the reader must accept the deviations in Le Vaillant's account. On these pages he presents himself as a natural researcher of nature who followed

[84] Le Vaillant (1795, I, p. 177).

[85] According to Van Gelder (1997, p. 95), however, until deep into the eighteenth century it was more common for travellers to start the story of their travels with an account of their youth.

[86] "On trouvera peut-être étrange que, pour donner la relation d'une Voyage résemment entrepris en Afrique, j'aie été forcé de me replier sur le passé, & de conduire mes Lecteurs dans l'Amérique mériodionale sur les premiers pas de mon enfance; j'ai cru qu'il ne seroit pas mal à propos de justifier, par les commencemens de ma vie, ma manière de voir, de penser & d'agir qui conservera toujours le goût du terroir, & qui, jugé peut-être avec sévérité, ne manqueroit pas de choquer ces esprits intolérans qui ne souffrent jamais sans humeur qu'on leur enlève leurs préjugés, & qu'on ose heurter de front les principes & les usages jusques là généralement adoptés" (Le Vaillant, 1790, "Préface", p. viii).

in the footsteps of his parents and started collecting at a very early age—the favourite eighteen-century scientific activity. In my view this self-projection is also inspired by literature. In Le Vaillant's account nature was his main teacher, as in the case of the young Emile of Rousseau's eponymous novel before him (1762). Of Emile Rousseau says: "In this way I learnt to walk in the wilds and I was almost born as a natural man [...] nature was my first teacher".[87] Little Le Vaillant started with a collection of insects, but this was eaten pins and all by his tame monkey;[88] so he then switched to collecting birds, a speciality that made him famous after his South African journey.

He says that his education explains his subsequent wanderlust and also justifies his outbursts against established science that follow in his travel account. From the outset he presents himself as an unconventional investigator of nature, cultivated by nature itself, who has no time for the systematics of established science. Moreover, his lack of scientific training is remedied by his familiarity with nature and nature research from an early age. His sound scientific objectivity was revealed at an early age. When he stayed in Paris for three years as a young man he "burned incense every day at the statue" of the eloquent Buffon, but was nevertheless unable to "forgive the scientist the exaggerations of the poet".[89] This was simultaneously aimed at Buffon's famous style as well as at elements in Buffon's *Histoire naturelle* which he found disputable.

However, while he presented himself as a person without training he was not relinquishing scientific authority. He repeatedly distances himself from the scholarly establishment,[90] but this does not mean that he also surrenders all means to be taken seriously scientifically. Instead, his lack of scientific background and of literary abilities[91] is turned into an advantage that must heighten the reliability of his statements. There is definitely a "rhetoric of reliability"[92] in his account.

---

[87] Rousseau, *Emile ou de l'éducation* (1762). For Emile too, nature is the most important teacher in the earliest days of his life: "il [Emile] prend ses leçons de la nature et non pas des homes" (he (Emile) takes his lessons from nature, not from people) (Rousseau, 1969, IV, p. 361). The difference, however, is that Le Vaillant also studies nature, which is not a subject of *Emile* [...]. There is a sketchy article about the relation between Le Vaillant and (among others) Rousseau by Boisacq (1993).

[88] Le Vaillant (1795, I, pp. 3–5). This choice may be an echo of the seventeenth-century research of Surinamese insects by Maria Sibylla Merian (1647–1717).

[89] Le Vaillant (1790, p. xvi).

[90] Le Vaillant (1795, I, pp. 71, 177–178, 208–209; II, p. 54).

[91] Le Vaillant (1795, I, p. 44).

[92] Gitelman (1991, p. 56).

Not having any knowledge of writing books,[93] like Le Vaillant, may be an advantage because one can then observe more accurately. One can observe facts that escape the systematics of science. Is nature itself not unsystematic?[94] The lack of a refined style does not have to be a deficiency either. Ever since Montaigne the lack of literary abilities had been taken as proof of reliability, because untrained people were less inclined to present a false image of reality.[95] Le Vaillant also emphasises autopsy time and again, especially in the case of phenomena his readers may find unbelievable: "I will repeat it ad nauseam: I have seen it, and the most eloquent writings or the most brilliant orations will never make me think otherwise".[96] However, the rhetoric of reliability remains confined to individual assertions, whereas the travel account as a whole has a conspicuously literary form.

Conclusion

For Le Vaillant the literary demands of an entertaining tale outweigh the conventions of scientific reliability and completeness. He says that his account is based on stories he usually told his friends. It lacks a plan, and to tell the truth one of his listeners could also have set it down.[97] The tension of the story is more important than its completeness. A "skilful author" makes sure that he keeps the descriptive sections, which Le Vaillant calls excursions, as short as possible in order not to interrupt the flow of the story too much.[98] In ethnographic descriptions he therefore thinks that supplements to his extensive description of the Gonaquas should suffice.

This literary pragmatism in the travel account can to a lesser extent also be discerned in other travel accounts at the end of the eighteenth century, such as those of Anders Sparrman. The dull, systematically constructed travel accounts of the type that Kolb wrote had gone out of fashion. The encyclopaedically arranged information of Kolb's travel account may have been easy to consult, but it was difficult to

93 Le Vaillant (1795, I, p. 177).
94 Le Vaillant (1795, I, p. 71).
95 Gitelman (1991, p. 75). See also chapter 4.
96 Le Vaillant (1795, I, p. 209).
97 Le Vaillant (1795, I, pp. 177–178).
98 Le Vaillant (1795, I, p. 44).

read. Writers of travel accounts at the end of the eighteenth century therefore switched from the systematic description of elements of the foreign world to a coherent tale of experiences collected in the course of the journey. The systematic approach was not abolished altogether, but was made subservient to an overarching narrative perspective.[99] We can see this very clearly in Le Vaillant.

Le Vaillant continued in this direction, but took it further than other authors of travel texts because he also elevated himself to the central figure of the story. He regularly drew attention to himself as traveller—"Me voilà"[100]—more than was necessary for the progress of the story. For instance, he provides a detailed description of the finery he dons for a visit to the chief of the Gonaquas. In Le Vaillant's representation this finery served a function in the interaction with the Gonaqua chief, because he could show in this way that he was different from the trekboere of whom the Gonaquas had been the victims. Whereas the trekboere were dressed slovenly, Le Vaillant was extravagantly decked out and sported a (beautifully combed) beard. The description of his appearance is at the same time intended to entertain the reader, in the same way as the anecdotes about his baboon Kees. With a dig at the scholarly travel accounts, he says of the Kees episode: "These details [of the behaviour of Kees, the baboon] may appear insignificant [...]" but may be more interesting than extensive and boring zoological descriptions.[101]

The overall impression left by Le Vaillant's work is that the representation of reality had been subordinated to a discursive presentation intended to sweep the reader along. For this reason an anonymous German reviewer of the *Voyage*, writing for the natural science journal *Magazin für das Neueste aus der Physik und Naturgeschiche*, was of the opinion that Le Vaillant's travel account excelled more through "Unterhaltung" (entertainment) than through "Gelehrsamkeit" (erudition).[102] Le Vaillant provides this entertainment not only by telling amusing stories, but also by connecting to the world of his readers' ideas. Rousseau and his

[99] The German literary historian Hentschel (1999, p. 37) even refers in this case to a paradigm shift in the poetics of the travel account at the end of the eighteenth century. Contarini (2004, p. 266) sees the novelty of Le Vaillant's travel account in the combination of an autobiography and a description of nature. However, she forgets that this had already been done by Sparrman. Le Vaillant simply continued this trend.

[100] Le Vaillant (1795, I, p. 134).

[101] Le Vaillant (1795, I, p. 177).

[102] "Anzeige neuer Schriften und Auszüge" (1791, p. 100).

primitivist precursors taught these readers among other things that savages were happy, and Le Vaillant only gave them what they wanted to hear. It is therefore not clear whether his representation of the Khoikoi is an apology or a way of pleasing his readers. Whatever the case, it had consequences for the quality of his ethnography. By modelling the Gonaquas on the prevailing primitivist and classicist schemes he may make them attractive to eighteenth-century readers, but it also takes away their uniqueness. Narina is not called Narina ('flower'), nor is she a Venus de' Medici.[103] The fact that it could also be done differently is demonstrated by the ethnographic descriptions produced by Gordon. Moreover, Le Vaillant's ethnographic discourse is contradictory because he alternates an indignation inspired by the literature at the impending disappearance of happy indigenous societies with colonisation plans.

From a scientific perspective as well, the emphasis on "Unterhaltung" was dangerous, because it detracted from his credibility. He extricated himself from this dilemma more or less by turning himself into a natural and Creole researcher of nature who by virtue of his familiarity with exotic nature considered himself elevated above the rules of science. Unlike European armchair scholars he regarded himself as not being bound by systematics. An exotic identity had to guarantee a superior understanding of exotic nature, while his unconventional narrative style made the information optimally accessible to his readers. In view of Barrow's reactions, this did not convince all of them, but other readers might feel that new observations compensated for this lack of authority.[104]

Although Le Vaillant's work was important for the poetics of travel literature, he had only a limited influence on the travel writing about South Africa.[105] His contemporaries in particular reacted against him.

---

[103] Strother (1999, p. 18) makes a comparable remark about the representation of 'Narina' by Le Vaillant: "Le Vaillant has stripped the figure of her true name and all the signs of Hottentot identity: hat, *kaross*, face painting, walking stick, or sandals. [...] Narina's image demonstrates that Le Vaillant can only transform the "Hottentot" into the noble savage by *evacuating all signs of the Hottentot*" (original italicisation). In my opinion Strother goes a little too far here, as he uses an archetypal Khoikhoi constructed by himself as a basis; the Gonaquas cannot have corresponded completely to this archetype because they had been subjected to strong Xhosa influences. But I do agree with Strother's main idea.

[104] The reviewer of the *Magazin für das Neueste aus der Physik und Naturgeschichte* (1791) did think that Le Vaillant had made any new discoveries.

[105] However, an anonymous German poet sang the praises of the meeting between Le Vaillant and Narina ("Vaillant und Narina", 1791, pp. 391–394). Owing to the wide

The only person who may have imitated him was the Dutch-German traveller Jacob Haafner (1755–1809). According to references in Haafner's *Verhandeling over het nut der zendelingen en zendelings-genootschappen* (Treatise on the utility of missionaries and missionary societies) of 1805, Haafner was familiar with Le Vaillant's travel accounts.[106]

It is therefore quite possible that the description of a romance between Haafner and "a young Hottentot girl" in his posthumously (1820) published *Lotgevallen en vroegere zeereizen* (Vicissitudes and earlier voyages) owes its existence to Le Vaillant's Narina story.[107] Haafner says that he met the girl during his second stay at the Cape (1769–1770) when he was looking for plants on the 'Lion's Tail' mountain near Cape Town. He saw her on top of the mountain. She turned out to belong to a group of independent Khoikhoi—like Narina—that was visiting the Cape to negotiate with the Dutch administration. At close quarters, this anonymous girl also was pretty:

> Apart from the curious mixture of fat and soot with which she was smeared, like all others of her nation, and which actually did not detract anything from her figure, she was the prettiest and most attractively shaped female figure one can possibly imagine.[108]

She contrasted strongly with the "usual hideousness of her nation",[109] which Haafner elaborates on in a footnote. Besides the girl remaining anonymous, the meeting is also described much more soberly than the Narina episode in Le Vaillant. Nothing much more happens than that the girl strokes Haafner's face with her hand. When he looks for her the next morning, her tribe has disappeared. On the mountaintop he finds only the beads she had worn: "This was the only memento I

distribution of the *Voyage* there are probably more literary adaptations of this episode. A South American parallel of Le Vaillant's Narina story is that of John Gabriel Stedman (1747–1797), who had lived in Surinam, about his love for the Negress Johanna in his *Narrative of a five years' expedition, against the revolted negroes of Surinam*, of which the first English edition appeared in 1796 (Stedman sa), the French translation in 1797 and a Dutch translation in 1799–1800. Le Vaillant cannot have known this story, however, because his *Voyage* appeared before Stedman's book (with thanks to Jean Kommers). The opposite is more likely, especially in view of the great popularity of Le Vaillant in the nineties of the eighteenth century.

[106] Haafner (1993, pp. 39–40).

[107] Haafner (1992).

[108] Haafner (1992, I, p. 78).

[109] Haafner (1992, I, p. 78).

had of my Hottentot love".[110] The illustration to Haafner's love story (fig. 32) suggests more than the text and seems a curious mixture of Le Vaillant's picture of Narina (fig. 30) and that of the woman who opens her karos in order to reveal her Hottentot apron (fig. 3). Of the latter there is nothing visible here either, because the girl is wearing a primitive fantasy skirt around the waist. New, of course, is that Haafner is shown with a walking stick in his hand next to the girl and that the Dutch flag is flying above the couple.[111] When Haafner's book was published in 1820 this flying flag was merely a memory. In the meantime the English had taken possession of the Cape Colony. From that time, English travellers began to have a decisive influence on the image projected of South Africa.

[110] Haafner (1992, I, p. 80). The editors of Haafner's work, De Moor and Van der Velde, establish no connection between Le Vaillant and Haafner; neither does anyone else who has published anything about Haafner.

[111] This picture has recently become popular as an illustration of miscegenation in the Cape Colony. See among others the cover illustration of the second edition of Heese (2005).

CHAPTER SEVEN

A 'BLACK LEGEND' OF DUTCH COLONIALISM IN THE *TRAVELS* (1801–1804) OF JOHN BARROW

Ethnographic curiosity in the eighteenth century was not limited to exotic peoples. In Europe a well-established and popular tradition of typologies of European peoples existed in which specific characteristics were ascribed to each people: Spaniards were proud, Germans drank too much, the French were frivolous and 'Muscovians' were barbarians.[1] Like in ethnography of people outside Europe, physical abnormalities also aroused interest. For example, all inhabitants of Styria in Austria suffered from goitre. In some women this swelling was said to be so large that they had to fling it over their shoulder in order to breast-feed their children. This peculiarity was as persistent in the description of the women of Styria as the 'Hottentot apron' in the ethnography of the Khoikhoi. In the nineteenth century Charles Dickens still used this gigantic tumescence in *The Uncommercial Traveller*.[2]

Travellers in the second half of the eighteenth century therefore also took an interest in the South African colonists. Peter Kolb described the colonists in the third part of *Capvt Bonae Spei Hodiernvm*.[3] Among other things he described their wedding and funeral practices and the conflict between the freeburgers and governor Willem Adriaan van der Stel in 1706. Later travellers focused more on the role of the colonists within the colonial dispensation the travellers were criticising. As independent ethnographic objects of study they were less interesting to these travellers because their novelty value was less than that of indigenous peoples. This was probably why the part of *Capvt Bonae Spei Hodiernvm* in which Kolb described the colonists was not translated into English and the French translation of this part was not reprinted.

The cursory assessment of the colonists by the 'philosophical' travellers was generally unfavourable. Sparrman thought that the lazy

---

[1] See among others: Meijer Drees (1997); Stanzel (1999); Koolhaas-Grosfeld (2003).
[2] Stanzel (1999, p. 15).
[3] Kolb (1719, pp. 583–846).

colonists wanted to make slaves of the Bushmen.[4] Le Vaillant held that the colonists were guilty of conflicts with the indigenous peoples and the destruction of the happy societies living in the interior which had not been corrupted by the decadent European civilisation.[5] Both Sparrman and Le Vaillant saw the threat to the indigenous population posed by the colonists in a broader context. Sparrman also criticised the Dutch colonial system in the Cape Colony,[6] with its unnecessarily cruel criminal justice procedures.[7] In Le Vaillant's view the pillaging the Spaniards had carried on in America in the sixteenth century and the cruelties in the Cape Colony at the end of the eighteenth century all lay on the same line.[8]

In 1801 and 1804 this negative picture was brought into sharper focus by the travel accounts of the English colonial official John Barrow (1764–1848; fig. 33). Barrow lived at the Cape during the first British occupation of the Cape (1795–1803). Unlike the Swede and the Frenchman, who came from countries that laid no claim to the Cape Colony, Barrow's observations on the Dutch colonists also have a political meaning. At the same time they are expressions of a new humanitarian sensitivity and aim to influence English colonial policy. In this chapter I want to investigate how Barrow merges the representation of the Dutch colonists and the Dutch colonial administration into a plea for maintaining or restoring British colonial authority over the Cape Colony after 1803, when the Cape reverted to Dutch administration.

Barrow's critical description of the colonists is based on an eighteenth-century humanitarian discourse and on arguments of natural law concerning the use of land in colonised countries. On this basis his arguments lead to an indictment of the Boer (the Dutch colonist or "boor", in Barrow's terms). In the first part of his travel account, *An account of travels into the interior of Southern Africa, in the years 1797 and 1798*, which appeared in 1801, he did this in the form of "sketches" with which he wanted to present a nuanced image of South Africa. Here the Boers were criticised only in passing. In the second part, *An account of travels into the interior of Southern Africa: in which is considered the importance of the Cape of Good Hope to the different European powers, as a*

[4] Sparrman (1977, pp. 131, 200).
[5] Le Vaillant (1790, pp. 146, 245, 331; 1795, II, p. 206).
[6] Sparrman (1977, p. 112).
[7] Sparrman (1977, pp. 49, 86).
[8] Le Vaillant (1790, p. 257).

naval and military station, published in 1804, their reprehensibility was emphasised more strongly in order to allow him to use this, in addition to arguments of military strategy and economic arguments, as a plea for recapturing the Cape Colony. In this chapter I want to examine how Barrow constructs his representation of the Boer and refer to the contrasting picture he presents of the indigenous population. The representations of Boer and native are related. Where some natives reached the highest steps of development attainable by barbaric peoples in Barrow's ethnographic discourse, the Boers were on the lowest level of European civilisation. From his point of view, degeneration was the opposite pole of the promise of development. This image went hand in hand with an anti-Dutch discourse. Barrow's degenerate Dutch colonist in the Cape Colony was the most conspicuous element in the negative view he and other English authors took of Dutch colonialism. In this way they constructed a 'Black Legend' of Dutch colonialism, as had been done in the sixteenth and seventeenth centuries with regard to the cruel Spaniards in South America.

Barrow's negative picture of the Dutch colonist in South Africa proved to be highly influential. Certainly in the first half of the nineteenth century authors subscribed to or rejected Barrow's opinion of the Boers,[9] and even at the end of the nineteenth century Barrow's opinions were reiterated without question.[10]

Barrow's career and modus operandi

It is worth noting that Barrow did not profoundly influence the image of southern Africa only. His *Travels in China*, which appeared in 1804 shortly after his books on South Africa, was of the greatest importance to the European image of China in the nineteenth century. Barrow was an early supporter of the stagnation thesis with respect to China. According to this thesis, China had a highly developed civilisation for centuries, after which development stagnated somewhere in the fifteenth century. By the end of the eighteenth century, the country had consequently fallen far behind technologically progressive Western Europe.

---

[9] Streak (1974).
[10] Van Wyk Smith (2003).

This thesis was the main presumption of the work of experts on China in the nineteenth century.[11]

In view of his humble origins, it is surprising that Barrow was able to become so influential. John Barrow was born near Ulverston in Lancashire in 1764 as the son of a small-scale farmer. His schooling was limited to a number of years in grammar schools, where up to his thirteenth year he received a thorough literary education in particular: Homer, Xenophon, Livy, Horace, Virgil, memorising large chunks of Shakespeare.[12] This shows up in his writings; in the books on South Africa he regularly quotes from English and Latin literature. In his spare time he also studied mathematics, navigation and surveying. After his school years Barrow became a clerk at an iron foundry in Liverpool. He then sailed to Greenland on a whaling ship, which gave him a lifelong interest in the Arctic. Upon his return he was appointed as a teacher of mathematics at a naval academy in Greenwich. One of his pupils was the son of the influential George Staunton, who introduced Barrow to George Macartney (1737–1806), a former dignitary of the East India Company. From this time on Barrow enjoyed the protection of influential Tories such as George Macartney.

In 1792 he accompanied Macartney on a two-year mission to China. On the journey back Barrow walked hundreds of kilometres across the country, during which he learnt a little Chinese and collected information about China that would be used by Staunton in his *Authentic account* [...] *of an embassy from the king of Great Britain to China* (1797). When Macartney was appointed governor of the Cape Colony, Barrow accompanied him as his secretary. From 1797 until 1803 he remained at the Cape, where he married Anna Maria Truter, the daughter of a Dutch official, in 1799.

Back in England he was appointed second secretary to the Admiralty in 1804, a post that he was to hold for almost 40 years. Here Barrow became an influential figure. In 1804 he published the second part of this book on South Africa, followed by *Travels in China* (1804) and *A voyage to Cochinchina* (1806) on his visit to Vietnam. At his instigation expeditions were sent to the Arctic in the second and third decades of the nineteenth century, and in Canada and Alaska Point Barrow, Cape Barrow, Barrow's Strait and Barrow Bay were named after him.

---

[11] Adas (1989, pp. 177–198).

[12] Barrow (1847, p. 6).

He published a book about polar journeys, *A chronological history of voyages into the Arctic region* (1818) and about the mutiny on the Bounty, *The eventful history of the mutiny of HMS Bounty* (1831). In 1830 he was one of the founders of the Royal Geographical Society. 1847 saw the publication of his *Autobiographical memoir*, a description of his career.[13] After his death a memorial tower was built on a field near Ulverston in his honour. However, this tower was unable to perpetuate Barrow's fame. It has been overtaken by the world of film, because according to a tourist website Stan Laurel, of Laurel & Hardy fame, who was born in Ulverston, is nowadays the best-known son of this little town.[14]

During his stay in South Africa Barrow made three journeys into the interior. The first journey, from July 1797 to January 1798, was undertaken at the request of Governor Macartney. On this journey to the Eastern Cape he visited Graaff-Reinet, Algoa Bay and the Winterhoek Mountains and conducted talks with the Xhosa chief Gaika (Ngqika). Barrow's brief was to gauge the political situation and the economic potential of the interior and to collect cartographic information. Apart from this, Macartney left Barrow at liberty; he relied on Barrow's curiosity.[15] From April to May 1798 he went on a second, shorter journey to Saldanha Bay and the Roggeveld and from late March to 8 June 1799 he accompanied a military detachment to refractory farmers in Outeniqualand and the region around Mossel Bay. The first two journeys are described in the first part of his books on South Africa, *An account of travels into the interior of Southern Africa, in the years 1797 and 1798* (1801). The account of the third journey appeared in chapter 2 of the second part, *An account of travels into the interior of Southern Africa* [...] *Volume the second* (1804).

Barrow had a limited amount of scientific equipment at his disposal on his travels. He had surveying instruments—a compass, sextant, surveyor's chain and chronometer—and a thermometer.[16] For identifying plants and animals he had Linnaeus's *Systema Naturae* (to which he refers repeatedly in his books) and *Hortus Kewensis* of 1789 by William Aiton.[17] In his autobiography he states that he originally spoke only a

[13] Barrow (1847); Lloyd (1970).
[14] www.lhdepot.com/Digest/Story.
[15] Theal (1898, pp. 110–1). Barrow wrote a report of his findings in 1798, the "General Description", which formed the basis of his *Travels* (1801 and 1804) (Boucher and Penn 1992, p. 91). For Barrow's work as a cartographer see Penn (1993).
[16] Barrow (1804, pp. 22–23).
[17] Barrow (1847, p. 143).

little Dutch,[18] which of course complicated communication with interpreters, colonists and the indigenous population. After his wedding to the Dutch Anna Truter in 1799 his knowledge of Dutch must have improved somewhat, but by then he had already completed his most important journeys.

Barrow collected information across a broad spectrum. Everything interested him: the three kingdoms of nature as well as the people he met in the interior. His contemporaries were struck by this all-encompassing scientific curiosity.[19] Governor Macartney wrote as follows to general Dundas about Barrow's journey to the Eastern Cape: "I have sent [...] Mr. Barrow, a Gentleman of my family, who is well qualified to observe, to judge, and to act, and whose journey, as he is known to be fond of natural history, passes for a tour, not of business, but of curiosity, science and botanical research."[20] The Cape society lady Anne Barnard describes Barrow as follows in her letters: "just one of the pleasantest, best-informed and most eager-minded young men in the world about everything curious or worth attention". He found "novelties and amusements" everywhere, "darted at plants and fossils" and "wherever we saw a questionable stone or ore, Mr. Barrow attacked it with a hammer".[21] Barrow had an all-encompassing thirst for knowledge, which was also typical of collectors in the early modern period.[22]

Before he published it, Barrow compared the information he collected with the existing literature. He had read the most important books on South Africa. The oldest authors—Tachard, Mercklin, Valentyn and Kolb—were, in his view, not competent because they had never visited the interior. Of the contemporaneous authors, Sparrman had good descriptions of nature, but he too readily believed Kolb's "absurd stories". Thunberg's book was "a collection of incomplete and unconnected paragraphs" and Paterson wrote "a mere journal of recurrences". The travel journals of VOC travellers, such as the journal the surveyor Brink kept of Hendrik Hop's journey in 1761–1762, were no good either. Le Vaillant's books contained valuable information, but he interspersed it with fiction. After all, what benefit could mankind possibly derive

[18] Barrow (1847, p. 157).
[19] See Daston and Park (1998, pp. 303–328) for a discussion of scientific curiosity in the eighteenth century.
[20] Theal (1898, p. 113).
[21] Barnard (1973, p. 46).
[22] Pomian (1990, pp. 43–60).

from Le Vaillant's amusing descriptions of his pet baboon Kees? It was quite annoying that "Monsieur" Le Vaillant was always the hero of his own tales.[23] In short, the existing literature was inadequate. The Dutch, being the colonial authority, had also neglected to describe and map the southern tip of Africa properly.[24] Sufficient reason to do it better oneself.

Barrow frequently confronted what had been said about a subject in scientific literature, especially as regards natural history, with his own experiences and hypotheses. He was a critical observer who compared existing knowledge with what he had seen himself and who sometimes proposed hypotheses that cast doubt on scientific paradigms. I will give only one example of this here. On his way to the eastern border of the Cape Colony he saw gnus. According to Linnaeus' *Systema Naturae* these animals belonged to the antelopes, which Barrow found odd because in his view a gnu had characteristics in common with the horse, the wild ox, the deer and the antelope. This suggested to him that there might be a need to introduce intermediate genera, i.e. animals that belong to more than one kind, into Linnaeus' taxonomy. This would at the same time show that the "grand chain of being" (Barrow) consisted of more intricate links than had been assumed until then.[25] Unlike Gordon, for example, he was not satisfied merely to play the role of a humble collector of data, but dared to put forward speculative proposals in order to match scientific theory better with empirical reality.

Besides travel books Barrow also drew the first reliable map of the Cape Colony. His books and map acquired the same significance in the nineteenth century as Peter Kolb's *Capvt Bonae Spei Hodiernvm* had in the eighteenth century: it was a point of reference for current observations. In the nineteenth century you could not ignore Barrow, even if your opinion differed from his.

Barrow's imperialist eyes

Thanks to Mary Louise Pratt, Barrow is now enjoying a dubious reputation in postcolonial criticism. On two occasions she attacked Barrow

[23] Barrow (1804, pp. 14–21).
[24] Barrow (1801, p. 8; 1804, p. 13).
[25] Barrow (1801, pp. 259–260). Here Barrow implicitly echoes the thought that the forms of life are interrelated to form a 'great chain of being' (cf. Lovejoy, 1965).

venomously as an exponent of the colonialist practice of science. In 1986 "Scratches on the face of the Country; or, What Mr Barrow Saw in the Land of the Bushmen" was published. In somewhat modified form, this essay was subsequently included in Pratt's influential study on colonial travel accounts, *Imperial Eyes*.[26] In this study she presents a highly charged image of the first part of Barrow's *Travels* (1801). Among other things, she reproaches Barrow for his great interest in natural history. Pratt sees this as a malicious attempt to "minimize the human presence".[27] She says that Barrow ignores human presence as much as possible. Barrow's "traveling eye" produces "empty" landscapes of which the economic potential is assessed. Where the inhabitants, including the colonists, are mentioned it is only to indicate the extent to which they do not meet Barrow's criteria. He generalises and objectifies their manners and customs, especially those of the San, who she says, are his "chief ethnographic interest".[28]

This characterisation of Barrow's work is one-sided and in many respects incorrect. Barrow most certainly takes an interest in the indigenous population of South Africa, and moreover the San are not his chief ethnographic subject; he has more to say about the Xhosas and the Dutch colonists. Barrow's description of the different population groups is much more complex than can be inferred from Pratt's gratuitous discussion, in which she has ripped pieces out of Barrow's work in order to "illustrate"[29] her views of scientific travellers at the end of the eighteenth century. Her main thesis is that this type of traveller was interested mainly in the description of landscapes and nature. If the indigenous population is discussed at all, the description is limited to physical characteristics, which are then in most cases described in unfavourable terms.

A large part of Barrow's books does deal with South African nature, which is described through Barrow's "traveling eye" or about which he postulates scientific hypotheses off the cuff. However, this interest in natural history must be seen against the background of eighteenth-century epistemology. Barrow was writing at a time when specialisation did not yet exist in scientific descriptions of the non-European world. Like his predecessors, he tries to describe everything that catches his

[26] Pratt (1992, pp. 58–68).
[27] Pratt (1992, p. 59).
[28] Pratt (1992, pp. 57–67).
[29] Pratt (1992, p. 39).

eye or that he thinks might interest his readers—the great Paarl Rock or the gnu as well as the manners and customs of the colonists and indigenous peoples. When he describes people, he often does so in generalising terms, because that was the custom in the early modern period, also when Europeans were described.[30] There was no distinct hierarchy between the topics that came up in travel accounts. All aspects of the foreign world were worth equal attention, although it is true that Barrow devotes more pages to the habits and customs of a particular population group than to the habitats of specific plants and animals or to geological phenomena. Ethnography therefore most certainly plays a relatively important role in Barrow's books, partly because this affords him the opportunity to take a stand with regard to the question whether the Cape Colony should be an English colony. In his ethnographic descriptions Barrow adopts an English colonial, 'enlightened' viewpoint in which he presents the indigenous population favourably or at least takes their side against criticism, whereas he sketches a very negative picture of the Dutch colonists and the Dutch colonial administration. In as much as there is any distancing from foreign people ('othering') in the *Travels*, it is aimed primarily against fellow-Europeans. I will return to this later.

On his three journeys, Barrow came into contact with—to use his own terminology—Hottentots, Kaffers, Bosjesmans and Namaquas, in that order. He was also in constant contact with the Dutch colonists. In his two books this contact resulted in descriptions of these five groups, where Barrow has least to say about the 'Little' Namaquas in the north-west of South Africa because in his time there were few members of this group left.[31] Most of the Khoikhoi (Hottentots) in the eastern parts of the Cape Colony had also lost much of their original way of life, so that Barrow had little to report about them.[32] The traveller's silence on a particular indigenous population group could therefore be the result of their disappearance due to acculturation, something that escapes Pratt.[33]

Barrow frequently takes up arms against critics of indigenous groups. He admits that the Khoikhoi in the Cape Colony are in a wretched state, but it is unjust to condemn them for that, as the colonists have

---

30 Stanzel (1999); Koolhaas-Grosfeld (2003, p. 210).
31 The 'Great' Namaquas lived in the present-day Namibia.
32 Barrow (1801, p. 144).
33 Guelke and Guelke (2004, pp. 26–27) also level this criticism at Pratt.

robbed them of their means of subsistence;[34] it is true that the San (Bushmen) are wild savages, but they were reduced to this because the colonists stole their women and children—they used to be peaceful;[35] the badness of the Xhosas (Kaffers) was simply a malicious fabrication by the colonists.[36] He endeavours to emphasise the good characteristics of indigenous population groups, especially of the Xhosas he visited on his first journey to the eastern border of the Cape Colony. In fact, contrary to what Pratt asserts[37] it is the Xhosas, and not the San, who receive most of Barrow's attention. They were a more newsworthy subject; not much had been published about the Xhosas because Le Vaillant, in his *Voyage* and *Second Voyage*, had only reported extensively on the San and Khoikhoi. This may well have been why Barrow devotes much attention to ethnography rather than to landscape descriptions in the part that deals with his visit to the land on the other side of the Fish River, the home of the Xhosa. Of the 35 pages he spends on this region 29 are devoted to the Xhosas.[38] Contrary to what Pratt suggests, ethnography is definitely not a subordinate subject in the *Travels*.[39]

Barrow starts the part about his visit to 'Kafferland' by describing his meeting with the young 'king' Gaika (Ngqika; 1775–1829), with whom he conducts negotiations to end the hostilities between the Xhosas and the colonists. Although he uses the discriminating distinction between civilised and uncivilised peoples in the description that follows, he also expresses admiration for the Xhosas in general and for Gaika in particular—he mentions Gaika's graceful appearance and surprising intelligence for one so young and "for a Kaffer".[40] Because of their relatively low level of technological development the Xhosas fall into the 'uncivilised' category. For example, their weapons are unwieldy,

---

[34] Barrow (1801, p. 156).
[35] Barrow (1801, p. 285).
[36] Barrow (1801, p. 176).
[37] Pratt (1992, p. 63).
[38] Region on the other side of the Fish River (Barrow, 1801, pp. 190–225); on the Xhosas (Barrow, 1801, pp. 193–221). In the second part of his *Travels* Barrow gives a shorter description of a few pages following his third journey (II, pp. 166–7).
[39] According to Pratt, Barrow's travel account consists "overwhelmingly of landscape and nature description" (Pratt, 1992, p. 59). To illustrate this she then quotes a short description of the region at the Fish River, where Barrow arrives in 'Kafferland'. What Pratt does not mention, however, is that this short description of the landscape is followed by a long ethnographic description.
[40] Barrow (1801, p. 195).

their manual skills limited and they have never ventured onto the sea. With regard to the latter he quotes a few verses from Horace (Odes I, 3, stanza 3), in which seafaring is called a form of a recklessness.[41] Technologically the abilities of the Xhosas were therefore limited, but they were not wanting in intelligence and other excellent (especially physical) qualities. This emphasis on physical and more specifically the physiognomic characteristics is neither a personal idiosyncrasy of Barrow's nor meant to degrade the Xhosas to natural phenomena, but arises from an eighteenth-century discourse on the basis of which psychological traits could be read from bodies.[42] The Xhosas displayed anatomic and physiognomic perfection. Anatomically some men resembled the pinnacle of male beauty, the Farnese Hercules,[43] which Joseph Addison called one of the four most beautiful classical statues.[44] The Xhosa's strong body was, according to Barrow, a result of a healthy, simple way of life, because he was not exposed to the debilitating influences of civilisation—an argument reminiscent of primitivism. Physiognomically speaking the Xhosa occupied the highest place in the 'Great Chain of Being', the hierarchical chain of life forms with the Europeans at the top:

> Though black, or very nearly so, they have not one line of the African negro in the composition of their persons. The comparative anatomist might be a little perplexed in placing the skull of a kaffer in the chain, so ingeniously put together by him, comprehending all the links from the most perfect European to the Ourang-Outang, and thence all the monkey-tribe. The head of the Kaffer is not elongated and the occipital bones form nearly a semicircle; and a line from the forehead to the chin drawn over the nose is convex like that of most Europeans. In short had not nature bestowed upon him the dark-coloring principle that anatomists have discovered to be owing to a certain gelatinous fluid lying between the epidermis and the cuticle, he might have ranked among the first Europeans.[45]

---

[41] "Illi robur et aes triplex/ Circa pectus erat, qui fragilem truci/ Commisit pelago ratem/ Primus" (He must have had a heart protected by a shield of oak and triple armour who first entrusted his fragile vessel to the tempestuous sea) (Barrow, 1801, p. 211).

[42] Magli (1989).

[43] Barrow (1801, p. 169).

[44] Haskell and Penny (1981, p. 229).

[45] Barrow, 1801, p. 205. According to the *Oxford English Dictionary* (http://dictionary.oed.com/) 'cuticle' and 'epidermis' are synonyms.

In the eighteenth-century physiognomic discourse it was assumed that the body could provide information about the soul. In this way Barrow was hoping to supplement what little Gaika was able to tell him through a Dutch-speaking Khoikhoi interpreter. This discussion was impeded by translation problems,[46] forcing Barrow to use other semiotic means to penetrate to the mind of the Xhosas. So he tried to read their bodies. In the passage quoted above he does this by applying the 'facial angle' of the Dutch anatomist Petrus Camper (fig. 34).[47] The facial angle made it possible to compare human types. The facial angle was calculated by drawing a line from the prominent part of the forehead to the upper teeth. This line intersected a horizontal line from the bottom of the nose to the ear. At the point of intersection these lines formed an angle that differed for various types of humans and apes. For orang-utans the facial angle was 58°, for Angolese and Kalmuk 70° and for Europeans 80°.[48] The greater the facial angle, the higher the rank of the type in the hierarchy—which extended further than living people. Roman statues reached angles between 85 and 90° and Greek statues 100°. According to Barrow, the facial angle of the Xhosas must have been around 80°, the same as that of Europeans. From the quotation it appears that Barrow views the dark skin colour as a minor cosmetic blemish. Elsewhere he says that the negative connotation of a darker skin colour is based solely on prejudice ("the prejudice of colour"),[49] a view he shares with Camper. The application of the facial angle and awareness of colour prejudice were still quite new in England at the time Barrow wrote about it.[50]

Physiognomy and anatomy not only provided general indications of the position of the Xhosas in the human hierarchy, but also pointed to intellectual qualities of which only limited evidence could be found in the Xhosas' material culture. "The ancients were of the opinion that

---

[46] Barrow (1801, p. 214).

[47] Camper's most influential work was his *Verhandeling over het natuurlijk verschil der wezenstrekken in menschen van onderscheiden landaart en ouderdom* [...] (1791).

[48] Cf. Meijer (1999, p. 108).

[49] Barrow (1801, p. 168).

[50] Meijer (1999, p. 167). After 1780 skin colour had become a subordinate category in the specialised debate between Johann Friedrich Blumenbach, Immanuel Kant, Johan Gottfried Herder, Petrus Camper and Georg Forster about physical differences. On the other hand a light skin colour was generally regarded as prettier (Mazzolini, 2005). More value was attached to anatomic and physiognomic differences (Bindman, 2002, p. 178). Barrow probably took note of this debate from a distance; it took place mainly in Germany in the eighteenth century and only reached England after some delay.

the face was always the index of the mind. Modern physiognomists have gone a step further", Barrow says.[51] In physiognomy, external beauty meant internal goodness and intelligence.[52] Therefore the face of the Xhosa proved his ability to reach a higher cultural level, which indicated that he was easy to civilise. Although Barrow paid the Xhosas compliments, the system he used for it was discriminatory. European facial features and statues from classical antiquity determined the position of non-Europeans in the hierarchy.

The Dutch colonists

Of the colonists Barrow had a very different opinion. In the above description this opinion was already suggested where he held the colonists responsible for the negative picture of indigenous population groups. Barrow's characterisation of the colonists is generally very disparaging. His description of them creates the impression that one would rather be at the top of the uncivilised people with the possibility of making it into the civilised group, such as the Xhosas, than to be in the rear of the civilised vanguard, such as the Dutch colonists in South Africa, the 'boors'. In Barrow's view, the colonists had only one virtue, namely hospitality;[53] everything else he criticised.

In the first part of his *Travels* of 1801 he merely gives a cursory sketch of the colonists (fig. 35). Here systematic overviews were short.[54] The attack is launched in the second part of the *Travels* as part of a plea for the repossession of the Cape Colony, which at that moment (1804) was back in Dutch hands. The depravity of the Dutch colonists then grows into a statement that must be proven in the following pages:

> In the mean time I shall just observe, as a position to be proved hereafter, that the Dutch peasantry of the Cape are better fed, more indolent, more ignorant, and more brutal, than any set of men, bearing the reputation of being civilized, upon the face of the earth.[55]

---

[51] Barrow (1801, pp. 207–208). The best known of these 'modern physiognomists' was Johann Caspar Lavater (1741–1801), author of *Physiognomische Fragmente zur Beförderung der Menschenkenntnis und Menschenliebe* (1775–1778).

[52] Cf. Tytler (1982); Magli (1989); Tytler (1995).

[53] Barrow (1801, p. 83).

[54] Barrow (1801, pp. 76–84).

[55] Barrow (1804, p. 79).

The butt of his criticism proves to be one category of colonists in particular, namely the stock farmers or, as they are known in modern historiography, the trekboere.[56] He also criticises the residents of Cape Town—he calls them stupid, bloated drunkards and keepers of slaves who read no books—but it is the stock farmers in the interior who are criticised most harshly. That they are well fed, as he says in the quotation, is a euphemism. In many cases the colonists are obese. Because of their bulging stomachs the men have trouble dismounting from their horses,[57] and it was among them that he saw the fattest woman he had ever seen in his life.[58] This obesity is associated with the second characteristic mentioned in the quotation, namely the colonists' indolence. He regularly returns to this point. Laziness had economic and moral consequences. Among other things, it led to poor utilisation of the natural resources of South Africa. The Dutch farmers were not interested in improving agricultural methods[59] and had made no attempt to domesticate quaggas and buffalo.[60] He even thinks that this is a regular Dutch colonial shortcoming, for in Batavia the Dutch are also lazy and they have the Chinese do all the work.[61]

The accusation of laziness is more important in a colonial context that appears at first glance. The laziness of the Boers means not only that they do not come up to the new standards of industriousness that apply in the industrial society.[62] The poor use of natural resources also takes away the justification for colonial occupation. In natural law from the seventeenth century onward, as formulated by Francis Bacon, Hugo de Groot, John Locke and Emeric de Vattel, efficient land use was the principal argument for taking possession of land on which nomads were living. After all, the nomads made no use of the land at all and therefore this land was *terra nullius*—nobody's property. But then the new owner also had the *duty* to use the land optimally.[63] According to Barrow, the Dutch colonists—and the nomadic trekboere in particular—did not obey this rule. Barrow maintained that under an English

---

[56] Especially Barrow (1804, pp. 93–111, 401–410).
[57] Barrow (1804, pp. 37–38).
[58] Barrow (1801, pp. 123–124).
[59] Barrow (1801, pp. 67, 89).
[60] Barrow (1801, pp. 93, 130).
[61] Barrow (1801, pp. 93, 130).
[62] Cf. Coetzee (1988, pp. 28–31).
[63] Cf. Muthu (1999, p. 974); Tuck (1999, pp. 232–233, 183); Pagden (2003, pp. 182–184).

regime more could be expected:[64] "the spirit of improvement that has always actuated the minds of the English in all their possessions abroad, will no doubt shew itself in this place."[65] Barrow uses the argument of natural law concerning land use against the Dutch colonial competitor. The Dutch do not meet the current standards for colonial occupation; Englishmen colonise better than lazy Dutchmen.

Natural law also obliged colonists to treat the local population well.[66] This precept was emphasised even more at the end of the eighteenth century as part of a new 'humanitarian' discourse. In this discourse the 'man of feeling' was praised as someone who was sensitive to the suffering of others and thereby gave evidence of a virtuous nature. The humanitarian feelings focused on an increasing number of marginalised groups, such as criminals, slaves and the mentally ill, and led to criticism of forms of cruelty that used to be acceptable.[67] At the end of the eighteenth century educated Englishmen also took issue with the maltreatment of animals.[68] Sympathy and an aversion to gratuitous cruelty were described as civilised emotions and such cruelty became a sign of barbarism. In the nineteenth century this humanitarian discourse developed into a special variant of the English colonial discourse. According to this humanely tinted colonial discourse an English colonial administration had to ensure humanitarian circumstances. The English public had to start identifying with the suffering of the oppressed Other, while the (cruel) colonists often became the target of criticism.[69] Opposition to slavery and sympathy with the maltreated native were presented as a special attribute of the British nation.[70] Barrow seems to have played a pioneering role in the development of this discourse, which only bloomed after his publications on South Africa in the nineteenth century.

The charge of cruelty against the indigenous people (and animals) plays an important part in Barrow's indictment of the Dutch colonists. It is also suggested in his argument that British authority could put an end to this. He constantly finds opportunities to provide examples

[64] Barrow (1801, pp. 8, 67).
[65] Barrow (1801, p. 17).
[66] Pagden (2003, p. 191).
[67] Cf. Spierenburg (1984, pp. 183–199); Laqueur (1989, pp. 176–204); Halttunen (1995, pp. 303–304).
[68] Turner (1980, p. 14).
[69] Lambert & Lester (2004).
[70] Lester (2001, p. 25).

of brutal acts by colonists against the indigenous population. He says that all the conflicts with the indigenous population are the colonists' fault,[71] and he records a large number of instances of maltreatment or murder by colonists.[72] "Inhumanity", he says, is the worst characteristic of the colonists.[73] One of these examples has become infamous: flogging of Khoikhoi where the duration was determined on the basis of the number of pipes that the punishing farmer smokes ("flogging by pipes").[74] Barrow introduces his description of this subject with the remark that in the Cape Colony one can find all the cruel punishments which are current in the West Indian slave colonies, but what he finds worst is the callousness with which the punishment is meted out. The Dutch farmer is the opposite of a civilised 'man of feeling':

> In offences of too small moment to stir up the phlegm of a Dutch peasant, the coolness and tranquillity displayed at the punishment of his slave or Hottentot is highly ridiculous, and at the same time indicative of a savage disposition to unfeeling cruelty lurking in his heart. He flogs them, not by any given number of lashes, but by time; and as they have no clocks nor substitutes for them capable of marking the smaller divisions of time, he has invented an excuse for the indulgence of one of his most favorite sensualities, by flogging them till he has smoked as many pipes of tobacco as he may judge the magnitude of the crime to deserve.[75]

Barrow gives examples of cruelty not only to people, but to animals as well. He calls the brutal way in which oxen were treated "an instance of brutality and cruelty that will scarcely be supposed to exist in a civilised country."[76]

Barrow paints a very dark picture of the Boers. If we run down the most commonly used *loci a persona*, as given by Cicero, the *Rhetorica ad Herennium* and in Quintilian, the Boers get very low marks on all points.[77] As regards the *body*, the Boers are too fat and too clumsy; their *character* is indolent, cruel and insensitive; *natural resources* are not well *used* as a result of indolence. Quintilian also mentions *descent* as an element of

---

[71] Barrow (1801, pp. 111, 165, 285).
[72] Barrow (1801, pp. 85, 145, 354); Barrow (1804, pp. 53, 96–97, 99–100, 133–134, 406–407).
[73] Barrow (1804, pp. 406–407).
[74] Barrow (1801, p. 146). Percival (1804, pp. 292–293) describes this punishment given to slaves.
[75] Barrow, 1801, pp. 145–146).
[76] Barrow (1801, p. 182); Barrow (1804, pp. 40–41).
[77] For the *loci a persona* see Lausberg (1960, pp. 131–134); Cicero (1949, pp. 343–345); *Rhetorica ad Herennium* (1954, pp. 175–183).

the description of a person.[78] Here again the Boers receive low marks. Their indolence indicated that they were not descended from the industrious Dutch.[79] Instead, they were probably descendants of German mercenaries who were used to maltreatment themselves, which also explains the cruelty of the Boers.[80] The result is a vituperation of the Dutch colonists, of which only the colonists of the Sneeuwberg region in the east of the Cape Colony, who are obedient to the English administration, are exempted.[81] In some cases this vituperation becomes criticism of the Dutch colonial dispensation, which is held responsible for lax government, neglect of the gathering of knowledge about the colony and a cruel system of criminal justice.[82] On two occasions Barrow even draws parallels with other Dutch colonial establishments. Lazy colonists are also found in Batavia[83] and flogging by pipes was, he said, also applied in the Dutch colony of Malacca.[84]

The remedy for these social evils was a strict but just English colonial administration. The Dutch administration before 1795 had neglected the colony. If the English had not come, the colony would have been given over to "Jacobinism",[85] a name for radical democratic politics inspired by the French Revolution and found at the Cape especially among the rebellious stock farmers of Graaff-Reinet, the same group that had to bear the brunt of Barrow's ire.[86] As long as the English were in charge at the Cape, Jacobinism stood no chance. The English administration had humanised the criminal law[87] and had to ensure that acts of cruelty committed against the indigenous population by the Boers were punished.[88] The underutilisation of land could also be improved under English authority.[89]

Barrow used a curious example to illustrate how good government could help prevent the brutal actions of the Boers.[90] During his stay on

[78] Quintilian (1958, pp. 468–469).
[79] Barrow (1804, p. 138).
[80] Barrow (1804, p. 101).
[81] Barrow (1801, p. 251).
[82] Barrow (1801, pp. 8, 17, 44–45).
[83] Barrow (1804, pp. 29–30).
[84] Barrow (1801, p. 147).
[85] Barrow (1801, p. 52).
[86] Philp (1998).
[87] Barrow (1801, pp. 44–45).
[88] Barrow (1804, p. 133).
[89] Barrow (1801, pp. 17, 67).
[90] Barrow (1801, pp. 268–275).

the eastern border of the Cape Colony he organised a punitive expedition ('commando') consisting of armed Boers against the San. The story of this commando is told in a style that differs from Barrow's neutrally formulated observations, as it is told as a personal experience with attention to the details of the events (the rhetorical figure *enargeia/demonstratio*). The objective of this expedition was twofold. For Barrow himself the aim was to see how the Boers acted against the San ("be eye-witness of the manner in which the farmers conducted expeditions against these miserable set of beings").[91] This was because he suspected the farmers of using unnecessary force against the San. However, to the farmers he told a different story. He ordered them to use force only in self-defence, as the sole object was to have a discussion with the San chiefs.[92] The double purpose makes this passage very ambiguous. The impression is created that he wanted to conduct an experiment—observing how Boers mistreat the San—where the participants in the experiment should not be aware of their role, while on the other hand he did not want to be an accomplice to crimes of the kind he abhorred himself. Hence the order to use force only in self-defence.

After the San had been located near the *Tower-berg*, the expedition attacks their camp early the next morning and shots are fired. When Barrow asks the leader of the commando why he shot despite the instruction not to, he answers: "Good God! [...] have you not seen a shower of arrows falling among us?"[93] But Barrow has seen no arrows, and encourages the Boers to look for them as proof—with the intention of giving the San the opportunity to escape, thus displaying humanitarian conduct. The experiment suffices, however, to see that for the farmers shooting San is a sport: "Like true sportsmen when game was sprung, they could not withhold their fire".[94] That the San had indeed shot arrows at them is something the reader only learns a few pages later, after Barrow has described how the San dare come closer and accept gifts, after the fright of the attack.[95] So the Boers had indeed fired in self-defence and not just for fun.

There is something insincere about the passage: the double objective, the temporary suppression of the justification of the farmers' fire and

[91] Barrow (1801, p. 269).
[92] Barrow (1801, p. 269)
[93] Barrow (1801, p. 272).
[94] Barrow (1801, p. 272).
[95] Barrow (1801, p. 274).

Barrow who directs and manipulates. However, the passage is also a demonstration of the humanitarian colonial policy that he advocates. The colonists were under his command, and force was thereby limited to a minimum and replaced with persuasion when the San were persuaded with gifts to stop their raids. According to Barrow these raids were a reaction to the crimes perpetrated by the colonists, who had driven the San off their land and abducted their children as slaves.[96] Nevertheless, Barrow ranked the San themselves lower in the hierarchy of peoples than the athletic Xhosas. They were a "disgusting pygmy race",[97] an opinion arrived at after having studied the infamous 'Hottentot apron' among the women.

A Dutch 'Black Legend'

Within the whole of Barrow's plea for retaining (before 1803) or recapturing (after 1803) the Cape Colony, his description of the colonists is but one of many arguments in favour of English rule of the Cape. The economic and especially the strategic military arguments set out in the second part of the *Travels* probably carried more weight. Whoever possessed the Cape controlled the sea route to Asia. Leaving the Cape in Dutch hands was dangerous, because in Barrow's opinion the Cape was in the grip of French agents during the period of the Batavian republic (1803–1806).[98]

With the negative picture of the Boers, however, Barrow also helped to create a Dutch Black Legend, a disapproving discourse about Dutch colonists and Dutch colonialism, which of course did correspond to reality in some respects.[99] This was also true of the original form of the Black Legend, the *Leyenda Negra*, the abhorrence cultivated by Protestant Europe of the cruelties the Spaniards had inflicted upon the Indians in

---

[96] For another reading of this passage, see Pratt (1992, pp. 66–67). My interpretation is concerned with the rhetorical aspects of this section (how Barrow wants to persuade the reader), while Pratt approaches it psychologically. Pratt makes the following remarks: "The venture seems to have been a traumatic one for him [Barrow] [...] outburst of emotion [...] Barrow's loss of innocence". Nevertheless, Pratt also points out the moral ambiguity of this passage.

[97] Barrow (1801, p. 279).

[98] Barrow (1804, pp. 139–140).

[99] See for example Newton-King (1999, pp. 105, 112, 116–149) and Penn (2005, pp. 146–148; also based on statements by Barrow).

South America, the main source of which was the frequently reprinted and translated *Brevísima relación de la destrucción de las Indias* (Short Account of the Destruction of the Indies) of 1552 by Bartolomé de las Casas.[100] In England this Black Legend was used in the seventeenth century to make English colonialism appear morally superior to the Spanish.[101] The Black Legend about Dutch South Africa displays conspicuous parallels to this, because accusations of cruelty and claims of moral superiority of British colonialism are both prominent in Barrow's books.

But he was not the only one. In 1804 *An account of the Cape of Good Hope* appeared, written by the English army officer Robert Percival (1765–1826). Percival had spent a few months at the Cape in 1796 and in 1801 and had travelled in the interior. Percival was in all respects Barrow's inferior. For a large part, his book repeats Barrow's accusations against the colonists and, more emphatically than Barrow's, against the Dutch colonial government. The Dutch colonists treat the indigenous population with cruelty,[102] they are dirty (p. 204), overweight (p. 205), indolent (pp. 205, 212–213, 220, 255), ignorant and stupid (pp. 206, 230, 280), ill-mannered (pp. 206, 277) and do not make proper use of the natural resources (pp. 211–242). As regards their descent the colonists are a mixture of adventurers of all shades, "low and profligate wretches" (pp. 4, 241–242). The brutal execution of slaves and black criminals is pinned on the Dutch government. Percival ascribes the degeneration of the colonists to the collapse of Dutch power. He says that the fall of the Netherlands is due to infighting. This has made it a mere shadow of what it used to be:

> The picture which Holland presents to all states of Europe, should be an awful, and will, I hope, prove a useful lesson to them. The degeneracy of the Dutch colonists ought indeed to surprise us less, when we observe the sad changes which have taken place in the spirit of the mother country, situated as it is in the midst of Europe, and of civilization. When we observe the present inhabitants of Holland, we can scarcely believe them to be the same people who formerly were so zealous in the cause of freedom.[103]

According to Percival the Dutch have become a sordid people, egoistic, incapable of intellectual effort, indolent and sluggish in their habits,

[100] Maltby (1971); Shaskan Bumas (2000).
[101] Shaskan Bumas (2000, p. 108).
[102] Percival (1804, pp. 92, 96).
[103] Percival (1804, p. 232).

without patriotism and, worst of all, they have betrayed their prince and allied themselves with the French and treated their good friends, the English, as enemies. One could therefore collectively call them Jacobites, to use Barrow's term. Because of this rotten attitude they neglect their colonies (pp. 233–238, 244). Similar sentiments were found in English colonial circles in Asia.[104]

The solution for the degenerated state of the Cape Colony is, as in Barrow, English colonial rule. After all, Percival says, it is a 'fact' that England never fails to introduce civilisation, abundance and industry to its colonies. The English would be able to uplift the indigenous population from their backward state, as witness the success they had with the detachment of Hottentot soldiers during the first English occupation (p. 93).[105] This was why the indigenous population preferred an English colonial authority (p. 250). Under English rule the colony would face a glorious future owing to the wise management by the British government, the efforts of the English colonists and because the indigenous inhabitants would under the benevolent policy "shake off their laziness". Slavery would be abolished in the Cape Colony simply because the English find slavery repugnant, although—not very consistently—it still remains a necessary evil in the English West Indian colonies (pp. 246–247).

In Barrow's and Percival's books it is not the indigenous inhabitants of South Africa, but the perverse Dutch colonialism (as they saw it) that is the target of criticism. By putting the conduct of the Dutch competitor and its local representatives, the colonists, in a bad light, English colonialism is justified, as was done in the sixteenth and seventeenth century by exploiting the Black Legend of Spanish crimes in South America. Postcolonial criticism readily overlooks these forms of 'othering' because it is generally preoccupied with the black/white issue and the representation of a coherent European colonial discourse—some form of 'orientalism'[106]—that is aimed exclusively at the non-Westerner. However, the European competition is also often looked at askance with an imperialist squint.

The image of the Boers created by Barrow (and Percival) had long-lasting consequences. The nineteenth century English were firmly

[104] Marshall (1988); Taylor (1988, pp. 126–147).

[105] A detachment of Khoisan soldiers already existed in the VOC period; Percival (1804, p. 93).

[106] Said (1991).

convinced that the Boers were "a backward, indolent and cruel of race of men".[107] One could, like Thomas Buxton, the leader of the abolitionists, make allegations in the English Parliament in 1834 on the authority of "Mr Barrow of the Admiralty" about the conduct of the Boers towards the indigenous population.[108] However, contemporaneous observers saw that many ills persisted under English colonial administration. The Scottish missionary John Philip (1775–1851), for example, wrote that when he came to South Africa in 1819 he at first believed, on Barrow's authority, that the evils were exclusively the fault of the Dutch colonial rulers, but that to his disappointment he had to admit that the new dispensation did not fare much better.[109] As we shall see, Barrow's representation of the South African colonists was initially believed in the Netherlands as well.

---

[107] Streak (1974, p. 121).

[108] Streak (1974, p. 7). According to Merians (2001, pp. 220–221) Barrow's perspective of the Hottentots and the Boers even became the official British opinion at the beginning of the nineteenth century.

[109] Philip (1828, I, p. xvi).

CHAPTER EIGHT

BATAVIAN COLONIAL POLITICS AND TRAVEL ACCOUNTS ABOUT SOUTH AFRICA

Early in 1803 the Cape Colony was returned to the Batavian Republic.[1] The Cape had been occupied by the English in 1795, but in 1802, with the treaty of Amiens, it was agreed to end the occupation. On 1 March 1803 commissioner-general Jacob Abraham de Mist (1747–1823) and governor Jan Willem Janssens (1762–1838) took over the government of the Cape Colony from the English. The Batavian regime did not last long. After hostilities had broken out in Europe, a British fleet of 61 ships sailed for the Cape. On 6 January 1806 an English expeditionary force landed at Blaauwbergstrand, north of Cape Town, where it engaged a Dutch force a few days later. The hastily assembled Batavian army consisted of a mixed lot of German mercenaries, Cape burgers, French marines, various Dutch detachments, 'Javanese' artillery and a Khoikhoi battalion, in short people "of all nations and languages [...] who had been assembled here, no less different from one another, starting with the most decent descendants of the original settlers to Eastern and Mozambican slaves".[2] Unfortunately, this multicultural little army was not efficient, but its composition did make it suitable for a commemoration of the battle in January 2006, in the twelfth year of existence of the multicultural 'new' South Africa. For this occasion the Batavian troops were elevated to the courageous defenders of the democratic 'rainbow nation' against a white British aggressor. In the description of this event special emphasis was put on the part played by the indigenous participants in the conflict, although if truth be told

[1] The Batavian Rupublic ('Bataafse Republiek') was the successor of the Republic of the United Provinces. It was created in 1795 after the French invasion of the United Provinces and it ended with the accession of Louis Bonaparte, the brother of Napoleon Bonaparte, to the throne of the Kingdom of Holland in 1806. During the Napoleonic Wars the Batavian Republic was an ally of France.

[2] In the words of governor Janssens, cited by Krynauw (1999, p. 85): "van alle taalen en landaard [...], die hier byeen gebragt waren, niet minder onderling verschillend, beginnende met de fatzoenlykste kinderen uit de volkplanting, tot Oostersche en Mozambicqsche slaaven incluzive".

many more European soldiers fell on the Batavian side.[3] The fact that this conflict in 1806 was actually a confrontation between two colonial powers was conveniently forgotten in the official commemoration.[4]

After the battle had been lost on 8 January 1806 governor Janssens signed the capitulation on 18 January on relatively favourable conditions. Janssens and the core of his troops were allowed to return to the Netherlands. However, the capitulation did not mean that the Netherlands waived its claims to the Cape colony; the Dutch recognised the English authority over the Cape only in 1814.

In the period during which the Batavian government was able to exercise power it was very diligent. The intention was to thoroughly reform the management and economy of the Cape Colony to the benefit of its residents and of the Netherlands. This also affected the production of documents. In order to establish their authority and familiarise themselves with the country, Batavian officials travelled into the interior and returned with accounts of their travels (fig. 36). In a little less than three years ten official travel accounts were produced.

With reference to these Batavian travel texts on South Africa, I want to show in the first place the extent to which texts of this type were related to the development of colonial policy. Their purpose was mainly to collect facts, and even when they were written with care they still had to serve the development of colonial policy. In principle, all knowledge about the Cape could serve this purpose, even if the author of a travel account held anticolonial viewpoints, as Sparrman did. In the second place I want to examine how the Batavian travel accounts can be distinguished from one another and what images of South Africa they project. As regards the latter I will confine myself mainly to the way in which the inhabitants are represented.

Most of the texts I will discuss were drawn up in the first place to contribute to the formation of colonial policy. Therefore they have more

---

[3] Krynauw (1999, pp. 84, 154). According to Couzens (2004, p. 45) 188 regular Dutch soldiers, 110 French marines, 4 colonists, 17 'coloureds', 10 'Malays' and 8 slaves were killed in action.

[4] The Cape daily *Die Burger* even published a special commemorative supplement on 6 January 2006. One of the articles has the meaningful title: "Moslems, boere en bruines veg dapper teen Britse invallers by Blaauwberg" (Muslims, Boers and Coloureds bravely fight British invaders at Blaauwberg). In South African historiography and in historical fiction the Batavian period has always been presented favourably. Because the Batavian period was short and the governors had so many good intentions it has always been a convenient screen for projecting historical wishful thinking.

in common than the works of scientific travellers do, but are at the same time less uniform than the accounts of VOC journeys; during the brief Batavian period no strongly developed 'centre of calculation' existed at the Cape. It can be assumed that the Batavian period was too short to formalise the collection of information, as the VOC had done. Instead, the educated Batavian travellers in particular followed—consciously or unconsciously—the published eighteenth-century travel literature.

Travel accounts and colonial policy-making

When a new colonial policy had to be developed for the Cape Colony with a view to its return by the English, the Dutch had to rely completely on travel accounts and the information that had been stored in the VOC archives; between 1795 and 1803 almost no contact had existed between the Netherlands and the Cape. When it seemed that the Cape would again be put under Dutch control, Jacobus Abraham de Mist, the later Batavian commissioner-general of the Cape Colony, was instructed to draw up a report on the way in which the administration of the Cape would have to be reformed.[5]

De Mist proposes several far-reaching reforms in his 1802 report. The times when the country was a failed colony were past. The Cape would be administered as a Dutch province; new districts would be proclaimed that would not have names, but numbers; the administrative apparatus would be put on a new footing, which among other things would allow for better financial control. As the interests of the colonists received priority, it would be made possible for them to trade freely with the Netherlands and the Dutch colonies. The aborigines were not overlooked either. The blot on the Dutch escutcheon due to reports in travel accounts about maltreatment of aborigines by colonists had to be wiped out by measures to protect the indigenous peoples.

What is remarkable about this report—which was accepted by the Aziatische Raad (Council for Asian Possessions and Establishments) of the Batavian Republic as the basis of a new policy for the Cape Colony in 1802—was that it was drafted by a person who had never been to South Africa. De Mist obtained his information from the old VOC archives and, for the period after 1795, especially from John Barrow's

[5] De Mist (1920).

Travels (1801). Travel accounts such as these had to bring the distant world closer and make it understandable, "ongeacht de menigvuldige, daarinvoorkomende dwaalingen, vergrootte voorstellen, en scherpe gispingen, aan nationaalen trots en naayver toeteschryven"[6] (despite their many misconceptions, exaggerated proposals and sharp censure attributable to national pride and jealousy). De Mist was aware of the shortcomings of his sources, but he had nothing better.[7]

He was not the only one who relied on the accounts of other people's travels for his design of a new colonial policy. After the Cape had been handed back to the Republic with the treaty of Amiens, the statesman of independent means Gysbert Karel van Hogendorp (1762–1834, who later became known as the leader of the Orangist movement of 1813 and the founder of a constitution for the Kingdom of the Netherlands) drafted his own version of a new colonial policy for the Cape.[8] He had not visited South Africa himself either and, unlike De Mist, he had no access to VOC documents. He therefore based his draft for a new colonial policy for the Cape mainly on published travel accounts. He presented his plea for the development of the Cape as an agricultural colony of which the products, as in De Mist's policy, could the traded freely within the Batavian commonwealth, as not much more than a systematic arrangement of the empirical information he had found in the travel accounts.[9] Ideally, the reader of his draft should start with the collection of fragments from travel writings he had put into the "Bylaagen" (annexes). In the body of the draft he regularly refers to these fragments in the annexes.[10]

What becomes apparent from the above is the enormous authority attributed to published travel accounts, despite their shortcomings. If in Europe one had no colonial archives at his disposal, one had to resort

---

[6] See "Rapport van het Departement tot de Indische zaaken" (Jeffreys & Gie, 1920, p. 4). Besides Barrow the travel accounts of Kolb(e), Damberger, Le Vaillant, Sparrman and Stavorinus are mentioned as sources of information for the period before 1795.

[7] De Mist (1920, pp. 34–35).

[8] Van Hogendorp had already drawn up an outline of a draft in his *Verhandelingen over den Oost-Indischen Handel* (Van Hogendorp, 1801, pp. 48, 93–100). After Amiens he elaborated his proposals for a colonial policy for the Cape in a separate book, *Verhandelingen over den Oost-Indischen Handel, Tweede Stuk. Behelzende eene staatkundige beschouwing van de kolonien benoorden en beöosten, Kaap de Goede Hoop. Met Bylaagen* (Van Hogendorp, 1802).

[9] Van Hogendorp (1802, "Inleiding", p. vii).

[10] Van Hogendorp (1802, "Inleiding", p. vi). The annexures ("Bylaagen"), numbered separately and totalling 371 pages, contain fragments from the main account of travels in South Africa in the eighteenth and the early nineteenth centuries.

to information collected in the accounts of travels. A critical attitude was adopted when reading this material, though. For example, Van Hogendorp draws a distinction between the different kinds of travellers and their interests:

> From all the known reports of the Cape of Good Hope, I have had to choose some and then again I had to confine myself to those that coincided with the perspective from which I view these colonies. The Botanist attaches great value to all kinds of plants that are relatively unimportant to Political Economy, while he mentions others that serve as food to the inhabitants or have substantial trading value. The Geologist pores over the mountain layers, without finding anything that is important to the Statesman, but he may also indicate the suitability of the soil for use, or for prospecting for useful minerals, or the position of mountains and rivers as defences against the enemy. I have therefore neither overlooked a single traveller nor used any one in its entirety as the basis of my work.[11]

By rearranging and reinterpreting the information of existing travel accounts it was made fit for purpose. Van Hogendorp called it "het gezichtspunt [...] uit welk ik deeze Kolonien beschouw" (the point of view from which I see these colonies). He presents the fragments from the travel accounts in a sequence that is suitable for his purpose and adds in brackets his own comments to the passages he quotes.

Although every text on South Africa had something to offer, there was no travel account that had a greater influence among the Batavians than that of John Barrow. The first part of his *An account of Travels into the Interior of Southern Africa* had appeared in 1801, shortly before the peace of Amiens. In the same year a French translation was published, and this was the version from which the Batavians obtained their information about Barrow's journey.[12] Barrow's map was the best that was

[11] "Onder alle de bekende Berigten van den Kaap de Goede Hoop, heb ik er eenigen moeten uitkiezen, en mij bij deezen nog bepaalen tot het geen in het gezigtspunt samentrof, uit welk ik deeze Kolonien beschouw. De Botanist hegt veel aan allerhande kruiden, die aan de Politijke Economie vrij onverschillig zijn; terwijl hij er anderen opnoemt, die of tot voedsel der Inwooneren dienen, of als een aanzienlijken handelstak den meesten aandagt waard zijn. De Geoloog zift over de lagen der bergen, zonder iets belangrijks voor den Staatsman te bepaalen; maar het gebeurt ook, de hoedaanigheeden van den grond ter ontginning, of ter ontdekking van nuttige delfstoffen, of de ligging van gebergten en rivieren ter verdediging tegen den vyand, op te geeven. Ik heb derhalven geen éénen Reiziger versmaad, noch ook een eenigen geheel tot grond van myn werk gelegd." Van Hogendorp (1802, "Inleiding", pp. iii–iv).

[12] *Voyage dans la partie méridionale de l'Afrique; fait dans les années 1797 et 1798* (Barrow, 1801). At the beginning of the nineteenth century few educated Dutchmen could read English.

available at the time; his book was the most recent one. It was also attractive because of the author's 'broad' perspective—Barrow had crossed the Cape Colony [met een] "Staat- Huishoud- en Natuurkundig oog [...] doorkruist"[13] (with the eyes of a statesman, an economist and a naturalist). Barrow's authority in the Netherlands was such that Batavian travellers were guided mainly by his book. To a considerable degree, one could read most of the Batavian travel accounts about South Africa as supplements to and corrections of the first part of Barrow's *Travels*.

Batavian accounts of travels in South Africa

After the Batavians had taken over the government from the English with much ado in 1803, information about the interior was urgently required. Owing to the vastness of the Cape Colony (as large as Spain, according to Van Hogendorp) and the poor accessibility of this newly acquired country it was difficult to assess the situation, especially in the fractious eastern part of the colony, from Cape Town. In order to gather the necessary information about this hinterland and to simultaneously firmly establish the new government up to the borders of the colony, various expeditions were undertaken, some of which yielded several travel journals:[14]

- The journey by governor Janssens to the Eastern Cape (2 April–13 August 1803) produced three accounts: an official acount written by Janssens' adjutant, captain W.B.E. Paravicini di Capelli, *Journaal en verhaal eener landreyse in den jaare 1803* (1),[15] a private diary kept by by Paravicini di Capelli containing more details than the official account, "Reize in de binnen-landen van Zuid-Africa" (2)[16] and lastly a journal by the colonist Dirk Gysbert van Reenen, "Dag-Verhaal

[13] De Mist (1920, p. 4).

[14] In the text below I refer to published versions of the travel accounts as far as possible.

[15] "Journaal en verhaal eener landreyse in den jaare 1803 door den gouverneur en generaal deeser colonie J.W. Janssens door de binnenlanden van Zuid-Africa gedaan" (Paravicini di Capelli, 1965). The author of this journal was captain W.B.E. Paravicini di Capelli, Janssens' adjutant.

[16] Paravicini di Capelli (1965, pp. 1–198).

eener Reize naar de binnenlanden van Afrika be-oosten de Kaap de Goede Hoop geleegen [...]" (3).[17]

- The round trip by commissioner-general De Mist through the colony (9 October 1803–20 March 1804) produced two accounts: first the official journal, written by A.L. de Mist, the son of commissioner-general De Mist, who accompanied his father as secretary: "Klad journaal der reize door den commissaris generaal Mr. J.A. de Mist en deszelfs gevolg in de binnenlanden van Afrika" (4);[18] then the private journal of commissioner-general De Mist himself: "Memoranda van de Kaapsche Landreize 1803" (5);[19] lastly the account by his daughter Augusta de Mist, "Dagverhaal van eene reis naar de Kaap de Goede Hoop, en in de binnenlanden van Afrika door Jonkvr. Augusta Uitenhage de Mist, in 1802 en 1803" (6).[20]
- The account of a journey by the surveyor J.W. Wernich to the Eastern Cape (1 April–24 July 1804): "Journaal of Dagregister der gedane reize over Zwellendam naar de Fleesch, Mossel en Plettenbergsbaaij [...]" (7)[21]
- The account of a journey by landdrost Van de Graaf(f) to the Tswanas ('Briquas') (7 May–7 August 1805): "Dagverhaal der Ryze van de Drosdye Tulbagh, naar de Briquas en terug" (8).[22]

---

[17] Van Reenen (1937).

[18] "Klad journaal der reize door den commissaris generaal Mr. J.A. de Mist en deszelfs gevolg in de binnenlanden van Afrika." Authored by A.L. de Mist, son of commissioner-general J.A. de Mist (Klad journaal, 1911).

[19] J.A. de Mist, "Memoranda van de Kaapsche Landreize 1803" (CA, Microfilm ZD/L4) (the journal ends with 17 March 1804).

[20] De Mist (1835, pp. 71–127). This was a fragment from a travel account published earlier in French for family and friends entitled *Rélation d'un voyage en Afrique et en Amérique par Madame* (De Mist/anonymous, 1821).

[21] "Journaal of Dagregister der gedane reize over Zwellendam naar de Fleesch, Mossel en Plettenbergsbaaij benevens een gedeelte van 't Auteniqua land als daar toe door den Gouverneur Generaal en chef Jan Willem Janssens benevens den Politiquen raade met den landdrost A.A. Faure gecommiteerd" (CA, BR 550, fol. 3–39). The author of this journal was J.W. Wernich (Blommaert, 1929, p. 44).

[22] "Dagverhaal der Ryze van de Drosdye Tulbagh, naar de Briquas en terug ter order van den Gouverneur en Generaal en Chef over de Colonie de Kaap de Goede Haap, en den ressorte van dien & Jan Willem Janssens, door den ondergetekende Hendrik van de Graaf, Landdrost van Tulbagh, gedaan in den Jaare 1805" (CA, BR 549, fol. 2–93). The author was the landdrost of Tulbagh, Hendrik van de Graaf.

- The account of the journey by governor Janssens to Saldanha Bay and St. Helena Bay (17–27 May 1804): "Journal eener landreize naar de Saldanha en St. Helena-Baayen in de Maand Mey 1804" (9).[23]
- The account by Dr Lichtenstein of a journey to the Karoo in order to persuade the residents of this area to have themselves inoculated against smallpox (19 August–9 September 1805): "Dagverhaal op eene reize door de Carroo" (10).[24]
- The account of the stock breeding commission to the Roggeveld and the Hantam district (25 September–9 November 1805): "Notul of Dagverhaal der Reis en Verrichtingen van President en gecommiteerde Leden uit de Commissie van Veeteelt [...]" (11).[25]

Lastly there is Baron Von Bouchenroeder's account of his journey to the eastern part of the Cape Colony. It was published in 1806 (12),[26] after his return to the Netherlands, and together with Lodewyk Alberti's *De Kaffers aan de Zuidkust van Afrika, Natuur- en Geschiedkundig beschreven* (1810, see chapter 9) it is the only travel account that appeared during the period the Netherlands laid claim to the Cape Colony. Except for the travel account by Augusta de Mist (1783–1832), the reports mentioned above have all survived as archival documents. Of the official documents a number appeared as historical source documents at the beginning of the twentieth century. The most comprehensive travel report that owes its existence to the Batavian period is *Reisen im südlichen Afrika* [...] (Travels in South Africa, 1811–1812) by the German doctor Hinrich Lichtenstein (1780–1857), who was initially a private tutor of Janssens' son and afterwards became major and officer of health in the Batavian army in the Cape Colony. Lichtenstein's *Reisen* and Augusta de Mist's travel account differ somewhat from the other administration-oriented Batavian travel accounts. Augusta de Mist wrote about personal experiences and Lichtensein wrote a scientific account that in many respects

[23] "Journal eener landreize naar de Saldanha en St. Helena-Baayen in de Maand Mey 1804" (Paravicini di Capelli, 1965, pp. 199–213).

[24] "Dagverhaal op eene reize door de Carroo" (CA, BR 549, fol. 96–132). The author was Hinrich Lichtenstein.

[25] "Notul of Dagverhaal der Reis en Verrichtingen van President en gecommiteerde Leden uit de Commissie van Veeteelt en Landbouw in de beide Roggevelden, den Hantam, &c." (Theal, 1911, pp. 324–435).

[26] *Reize in de binnenlanden van Zuid-Afrika, gedaan in den jare 1803* (Von Bouchenroeder, 1806).

emulated Barrow's *Travels*. He also included parts of Janssens' official travel report and De Mist's personal journal in his book.[27]

The administrative Batavian travel accounts can be divided roughly into two groups: journals that attempt to link up with an existing discourse on South Africa as established especially by published travel accounts written by a cosmopolitan group of travellers, and documents that do not. The authors of the first group, which for lack of a better term we will call 'erudite' travellers, are captain Paravicini di Capelli, commissioner-general J.A. de Mist, his son A.L. de Mist, Dr Lichtenstein and Baron Von Boucherroeder.[28] Their documents are characterised by a broad perspective on the South African world. The authors travelled across the Cape with the eyes of an economist and a naturalist and, as we will see, with attention to the landscape and its habitants. The other accounts were written with a much narrower objective and do not differ much from the journals kept by employees of the VOC until 1795. The information content therefore remained limited to matters that were directly related to the mission. For example, the report of the stock breeding commission of 1805 contains almost only information on the route followed and the efforts the commission made to persuade unwilling farmers in the Roggeveld to switch over to woolled sheep.

The limitation is especially striking in the account of landdrost Van de Graaf's journey to the Thlaping, the Tswana people (known in contemporaneous sources as 'Briquas') in 1805. At the point when Van de Graaf visited the Tswanas, contact between these Bantu-speaking people in the north-west of South Africa and the Cape Colony had existed for only a few years.[29] Consequently, much that was new could have been reported about them. However, Van de Graaf confined himself to the rather dry observation that the Tswana settlement he visited was quite large (some 600 houses, each with its own yard) and that the wide streets were full of people. On the same day he spoke to the 'king' of the Tswanas, but we do not learn more than that the talks concerned relations between the Tswanas and the missionaries.[30] When he had completed his duty, Van de Graaf promptly left.

---

[27] I will return to this later.

[28] The documents in question are 1, 2, 4, 5, 10 and 12.

[29] In 1801–1802, when the English ruled the Cape Colony, an expedition under Sommerville and Truter made a journey to the Tswanas (Bradlow, 1979).

[30] Van de Graaf (CA, BR 549, 30 June 1805, fol. 62–64).

Dr Lichtenstein, who accompanied him on the expedition, wrote a comprehensive and in some places even ecstatic report of the same visit to the Tswana settlement in his *Reisen im südlichen Afrika*. As he was walking through the village, he was overcome by "a most joyful feeling that fate had favoured me above so many others to walk among these remarkable people, whose existence as a more than half civilised people had remained unknown to science for so long and whom I came to love and esteem more highly by the hour".[31] The difference between Lichtenstein's and Van de Graaf's representations is that the former, as he himself says in the quotation, is written with scholarly objectives, as a traveller driven by curiosity. From this perspective every detail of the still unknown people was of the greatest importance and its discovery fired the enthusiasm of the travel writer. Van de Graaf deemed it sufficient that he had carried out his official duty in his report.

'Erudite' keepers of journals, such as captain Paravicini di Capelli and De Mist and son, followed in the tracks of the travel accounts that had been published in the last decade of the eighteenth and the beginning of the nineteenth century about southern Africa and frequently referred to them. The baggage Messrs De Mist and Lichtenstein took along on their journey to the interior contained all the authoritative books that had been published about South Africa.[32] Barrow's *Travels* occupied a special place; like a kind of *Lonely Planet*, it was constantly consulted and compared with their own observations. Sometimes even a mere reference to Barrow in their own text sufficed.[33] In other respects too the travelling Batavians wrote bearing in mind what they had read. This gave their own work a polished style and presentation, and especially a high information content.

But more important were the lessons in observation that the existing travel literature offered. From this they learnt that subjects of natural history, and especially the inhabitants, were worthy of interest and that one had to look further than merely from utilitarian, political and economical points of view. Especially in the personal travel journals of Paravicini di Capelli and commissioner-general De Mist, the view of

---

[31] "ein höchst freudiges Gefühl, dass es mir vom Schicksal vor so vielen Andern gestattet war, unter diesen merkwürdigen Menschen zu wandern, deren Dasein als eines mehr als halbgebildeten Volks so lange der Wissenschaft fremd geblieben war und die ich mit jeder Stunde lieber gewann und höher achten lernte" (Lichtenstein, 1812, II, p. 511).

[32] Lichtenstein (1811, I, p. 27).

[33] For example Paravicini di Capelli (1965, p. 78).

the world was very broad. A remarkable feature of the administrative journals is the attention paid to the beauty of nature, a subject that was never dealt with so extensively in earlier travel accounts. Landscapes were evaluated less in terms of their potential for agriculture and stock raising, as was customary in the 'non-erudite' accounts, and more in terms of its beauty. Pretty landscapes are "schilderachtig" (scenic), "picturesque"—which must often be taken literally: they are landscapes as in paintings. For example, De Mist (1803) notes in his personal journal for 16 October 1803: "Schoonen avond. Het ensemble van de Huismanswooning, stallen, slavenwooningen, vee, Paarden, Menschen enz. enz. maakte het schoonste tafereel voor het Penseel van Teniers, Wouwerman, van Breugel en dergelyke Boerelandgezichten-schilders."[34] (Beautiful evening. The ensemble of the homestead, stables, slaves' quarters, cattle, waters, people etc. etc. makes the most beautiful scene, fit for the brush of Teniers, Wouwerman, Van Bruegel and other painters of pastoral views.) Similar expressions, although somewhat toned down, were recorded in the official journal of his son, who accompanied him on the journey.[35] This fascination with the scenic landscape was a European phenomenon which has in English studies of the picturesque landscape been interpreted as an aristocratic perspective of the environment—the view of the landowner.[36] The Batavian travellers seem to have had a more purely aesthetic perspective.

De Mist senior appreciated green landscapes in particular, not the "horrible arid Karoo fields" (4 February 1804; fig. 37), although he found the "bold, lifeless" scenery with steep ravines and terrible mountains impressive (30 November 1803). Paravicini di Capelli was impressed by the steep chasms: "de woestheid van deeze diepte heeft iets prachtig schoon"[37] (the wildness of this depth has great beauty). And even the non-erudite surveyor Wernich once recorded his fascination with the environment. He says the following of the Indian Ocean surf at Mossel Bay:

---

[34] De Mist probably had the following painters in mind: David Teniers the Elder (1582–1649), Philip Wouwerman (1619–1668) and Pieter Brueghel the Elder (1525–1569). However, Wouwerman cannot truly be called a 'painter of pastoral scenes'; as he concentrated more on battle scenes and horses. De Mist probably mentioned him because of the latter subject.

[35] De Mist (1911, p. 126).

[36] Barrell (1988); Seymour (2000). For the picturesque landscape in nineteenth-century representations of Australia see Ryan (1996, pp. 60–76).

[37] Paravicini di Capelli (1965, p. 35).

> Then, being in this wonderful overhanging cliff or grotto, one can entertain oneself by watching the sea. At the foot thereof it rises up in violent agitation and forcefully drives the waves, which have rapidly built up, foaming over the rock. This was a spectacle that not only attracted my attention, but from which only the people with me and the falling dusk could tear me away.[38]

Father De Mist and son, Paravicini di Capelli (and, by exception, the surveyor Wernich) were therefore not insensitive to the sublime, wild landscape which educated Europeans had begun to appreciate by the end of the eighteenth century.[39] But like Paravicini di Capelli commissioner-general De Mist did feel more comfortable under shady trees near running water. He makes the following note in his diary about the farm De Vondeling:

> [...] *an entrancing location*, especially when on a hot day one comes out of the shadeless Africa and can for the first time sit down in a small copse of fine oak and orange trees bisected by a cool, running stream. Everything here resembles a beautiful farm in Gelderland. I will never forget this paradise [...] (21 October 1804; original emphasis.)[40]

Paravicini di Capelli made drawings of some of these 'scenic views'.[41]

The inhabitants

It is especially in the 'erudite' travel accounts that the inhabitants of the areas visited are described. The other accounts confined them-

---

[38] "In deze wonder overhangende rotze of grotte dan zynde heeft men 't vermaak de zee te beschouwen. Aan den voet derzelve zwelt zy met vreeslyke beweginge en drijft met drift de snel omhoog gereeze baaren schuimende over de rotze. Dit was een gezigt dat myn aandagt niet alleen naar zig trok maar waar ik niet dan door 't byhebbend gezelschap en den vallenden avond konde terug gebragt worden." "Journaal of Dagregister [...]", April 1804 (CA, BR 550, fol. 13–14, 27). However, this is the only time Wernich describes his experience of nature. The erudite travellers, on the other hand, do it repeatedly.

[39] Thomas (1984, pp. 254–269). The concept 'sublime' for wild landscapes that fill the viewer with fear and trepidation was developed by Edmund Burke (1729/30–1797) in *A philosophical Enquiry into the Origins of Our Idea of the Sublime and Beautiful* (1757).

[40] "[...] een *verrukkende ligging* vooral wanneer men op den heeten dag uit het onbelommerd Africa, *voor het eerst* zich onder een boschje van schoone eiken en lemoenboomen, met een koele loopende bron doorsneden, mag nederzetten. Alles hier naar een schoone Geldersche boerenplaats gelykende. Ik vergeet dit Paradys nooit [...]" (J.A. de Mist, "Memoranda van de Kaapsche Landreize 1803" (CA, Microfilm ZD/L4)).

[41] Paravicini di Capelli (1965, p. xxvi).

selves mainly to descriptions of the agricultural potential and the route followed.

Now it is true that route descriptions—with views (see above)—are not absent in the earlier 'erudite' journals either, but the focus is on the inhabitants rather than on the country. The only exception to this rule is the account by Von Bouchenroeder, *Reize in de binnenlanden van Zuid-Afrika* (1806). Judging from the introduction to his book, Von Bouchenroeder was quite familiar with the travel accounts on South Africa published earlier (and therefore also wrote 'eruditely'), but unlike his predecessors he pays minimal attention to the inhabitants. This is understandable, because Von Bouchenroeder wrote his account with a different aim in mind. Like Van Hogendorp he thought that the area around Plettenberg Bay was eminently suitable for establishing an agricultural colony. To this end he had come to the Cape with a group of colonists, and in 1803 he set out on a journey to the Eastern Cape to inspect this area. But for various reasons the colonial government abandoned the colonisation project, and after wrangling with Janssens and De Mist Von Bouchenroeder was banned from the colony. After his return to the Netherlands he published his *Reize in de binnenlanden van Zuid-Afrika*. It was an indictment of the Cape governors, whom he accused of capriciousness, and at the same time a substantiation of his view that his and Van Hogendorp's plans could certainly be realised.[42] To this end he supplies much information about the quality of the soil and almost none about the inhabitants of the regions he had visited.

But as said above, most of the 'erudite' travellers did take an interest in the population of the colony. In order to stimulate prosperity the government first had to get its affairs in order in the eastern part of the colony, where colonists had in the past rebelled against the VOC and the English and where rebellious Khoi, San and invading Xhosas threatened to destroy the hegemony of the Europeans. The main purpose of the journeys by governor Janssens and commissioner-general De Mist was to restore order in the border districts (fig. 38). Compared with these urgent problems, settlement projects and the improvement of soil utilisation—Von Bouchenroeder and Van Hogendorp's main goals—had no priority at all.

---

[42] For the colonisation project proposed by Van Hogendorp and Von Bouchenroeder see: Von Bouchenroeder (1806); Van Hogendorp (1876, pp. 211–225); Blommaert (1926); Van der Merwe (1926, pp. 285–292); Freund (1971, pp. 415–430).

Published travel accounts had a strong influence on the Batavian travellers' view of the local population. In this literature a definite image of the residents of the Cape Colony had crystallised by the end of the eighteenth century. Two criteria were applied for evaluating the residents: economic achievements (especially the labour capacity) and moral qualities. This gave rise to a contradictory hierarchy. In economic terms, the colonists and the 'Bastaards', descended from them by intermarriage with the Khoisan, were deemed superior. Morally, however, this group of Europeans and semi-Europeans was repugnant because of its cruelty against the indigenous peoples. The indigenous peoples were pitied because of the injustice that they had to suffer at the hands of the colonists, but frequently also admired—especially the Xhosas, who according to Barrow deserved a place alongside "the first Europeans".[43]

De Mist's own "Memorie" (memorandum) of 1802, written before he had seen South Africa with his own eyes, is in line with these current opinions, especially with Barrow's *Travels*. The "Memorie" presents an extremely negative picture of the colonists, particularly in the east of the colony. Their manners have been completely "bastardised" by the absence of intercourse with "civilised" people, their pastoral lifestyle and the constant consumption of meat. They hunt the San and Xhosas, whose life means nothing more to them than that of a hare or a wolf, on an organised basis. It is the colonists' fault that war is waged with the Xhosas. The difference between the colonists in the west of the Cape and these ruffians on the border is merely a matter of degree. In order to educate this bunch of Europeans gone wild, De Mist intends to launch a civilisation offensive. Churches must be built in order to help teach them manners, education must wipe out ignorance and see to it that even the Capetonians become Dutch again.[44] Similar opinions are also found in Van Hoogendorp.[45]

The low opinion De Mist had of the moral qualities of the colonists in the east of the Cape Colony is also expressed in a "Memorandum voor de Gouverneur Janssens over het karakter van de met name genoemde boeren" (Memorandum for governor Janssens concerning the character of the farmers mentioned by name) of 1803. In this memorandum De Mist states that Janssens must expect the border

[43] Barrow (1801, p. 206).
[44] De Mist (1920, pp. 44–50, 109–116).
[45] Van Hogendorp (1802, pp. 76, 84).

farmers to be "indolent, lazy and cowardly". He repeats that it is the misconduct of the farmers that is to blame for the war with the Xhosas and the Khoikhoi. The farmers are "incredibly stupid in their ideas, credulous, fearful, believing absolutely everything they are told and [...] [they] have the character of cowardly tyrants". They maltreat their servants. The document ends with a list of names and an indication of the qualities of the persons listed. Of the 50 farmers mentioned by name, 38 fall in the categories of "bad" and "scheming characters". Some are indicated more specifically as "a very bad case", "a tyrant" and even "an old beast". The aborigines, on the other hand, are good-natured folk who are merely conducting a war in reaction to crimes committed by the colonists.[46]

In keeping with the existing representation in eighteenth-century travel writing, it is argued in the 'erudite' travel accounts that the aborigines within and beyond the colonial borders (Khoikhoi and San) must be protected against the crimes committed by the colonists. To substantiate this, instances of maltreatment of Khoikhoi by colonists are mentioned repeatedly—"one can hardly take a step in this place without hearing a cry of maltreatment"—and explained negatively by, for example, the opinion "that these brutes [the colonists] stupidly imagine that they are higher, more privileged beings than the poor Hottentot who, because of his birth, must serve him".[47] And when a colonist called "Fereyra" explains to governor Janssens that the "Hottentots are of the race of Cham, cursed and doomed to slavery by God", the governor does not take this very well.[48]

His relatively long residence among the Xhosas enabled Paravicini to supplement Barrow's information in numerous respects, among others by compiling a better word list of the Xhosa language. He also describes the Xhosas particularly sympathetically, as in the following passage:

> [...] there are no more peaceful, accommodating people alive than this nation, always content, even if one rejects their begging. During our stay among the Xhosas we saw no sign of dissatisfaction, and we were truly sorry that we had to leave them so soon.[49]

---

46 "Memorandum voor Gouverneur Janssens" (Memorandum for Governor Janssens) (1932).

47 Paravicini di Capelli (1965, p. 34).

48 Paravicini di Capelli (1965, p. 49).

49 "[...] Vreedsamer, gedienstiger menschen als deze natie leven er niet; altyd tevreden, zelfs dan als men hunne bedelaryen afwysd. Wy hebben gedurende ons verblyf tusschen de Kaffers geen schyn van ongenoegen gehad, en het deed ons wezendlyk leed hun zoo spoedig te moeten verlaten." (Paravicini di Capelli, 1965, pp. 128, 140).

But while the representation of the indigenous population is fully aligned with the representations in authoritative travel accounts such as Barrow's, the Batavian travel texts do display a certain degree of balance in the representation of the colonists. On the one hand the misconduct of the colonists is recorded, but on the other hand certain positive traits are observed and the picture of the systematically abusive colonists is put into perspective to some extent. Thus Paravicini states that maltreatment is not the rule, which Barrow's *Travels* could have led one to believe, but occasional at best. Moreover, both he and De Mist senior express the opinion that Barrow's aversion to the colonists arose from xenophobia.[50] Among the positive characteristics of the colonists De Mist mentions their hospitality to travellers and the fact that they swear less than the Dutch.[51] His son finds it worth mentioning that among these people a woman can occasionally be found who would be accepted in good society.[52]

One explanation for this more balanced picture may lie in the political association which the Batavian travellers, unlike their non-Dutch predecessors, shared with the colonists: to them they were no strangers but "medeburgers"[53] (fellow citizens) of a distant province of the Batavian Republic. In Batavian usage, the term 'medeburger' did not refer primarily to political citizenship, to citizens as participants in the political process, but more to the citizenry as an object of administrative reform.[54] In the circumstances of the Cape Colony, improvements in the economy and in education and the granting of religious freedom aimed to create the circumstances for the development of competent and upright citizens who could grow the prosperity of the region through their own efforts. The Batavian policy was always aimed at improvement of the lot of the "ingezetenen" (residents), the colonists; the lot of the other inhabitants took second place. Nowhere did this become clearer than in De Mist senior's view of slavery. Although he finds slavery an abominable institution, it should for the sake of the

[50] Paravicini di Capelli (1965, p. 29); De Mist (1920, p. 112).

[51] "Memoranda van de Kaapsche Landreize 1803", 6 November 1803 (CA, Microfilm ZD/L4).

[52] De Mist (1911, 2 December 1803, p. 45).

[53] Freund (1971, p. 163).

[54] In the Dutch context, unlike in France, the concept of the enlightened citizen had mainly a moral content. For the ideas of the enlightened citizenship in the Netherlands around 1800 see: Kloek & Mijnhardt (2001, pp. 143–317); Kloek & Mijnhardt (2002, pp. 155–172).

colonial economy not be abolished on the spot.[55] The only important concession to the indigenous inhabitants that De Mist considers is shrinking the size of the colony, which was much too large, in order to give the San more space. In his view, this will put a stop to the war between the San and the colonists.

Taken all together, the Batavian travel accounts form a coherent colonial discourse about land use and inhabitants intended to assist the colonial government in getting a better grip on the interior (the details fall beyond the scope of this chapter). The travel accounts offered reliable information on the hotbeds in the interior such as the much-troubled eastern parts of the Cape Colony. But not everything in the accounts had been recorded for this purpose. The personal travel journals of De Mist senior and Paravicini di Capelli in particular contained much that went beyond political utility: accounts of troubles experienced on the journey, anecdotes about people encountered on the way, hunting tales and descriptions of landscapes.

Although the accounts were intended for official use, they could also become part of the general public discourse on southern Africa. This had happened with VOC texts before a number of times, as we have seen. In the case of the Batavian travel journals some texts were included in the *Reisen im südlichen Afrika* by Hinrich Lichtenstein, one of the most important and comprehensive scholarly accounts of travels in South Africa from the first half of the nineteenth century.[56] Lichtenstein had crossed the interior alone as well as in the company of De Mist and Van de Graaf. For his *Reisen* he made use of official and personal Batavian travel accounts,[57] especially the 'erudite' ones, which because of their greater wealth of information also had the most to offer. He adapted parts of Paravicini's official journal of governor Janssens' journey.[58] He also made extensive use of the two travel journals of De Mist senior and junior, whom he accompanied. The report of this journey forms the core of the first part of his book.[59]

Lichtenstein is an example of a scientific traveller who knew how to use the colonial system for his own books. He accompanied governor

---

[55] De Mist (1920, pp. 109–112).

[56] Translations of Lichtensteins *Reisen* were published in English (1812), Dutch (1813–1815 and 1818) and French (1842) (Lichtenstein, 1973, pp. 6–7).

[57] Lichtenstein had collected a large number of Batavian documents (Lichtenstein, 1973, pp. 85–88).

[58] Lichtenstein (1811, I, pp. 500–545; 1812, II, pp. 61–103).

[59] Lichtenstein (1811, I, p. xxxiii).

Janssens as his son's tutor, because this offered him the opportunity to realise a long-standing desire to visit Africa. In the Cape Colony he was able to undertake, alone and in the company of Janssens, journeys that supplied material for these travel books. His South African journey formed the basis of his illustrious career after his return to Germany. Because of his South African journey, Alexander von Humboldt (1769–1859) invited him in 1811—even before the publication of his *Reisen*—to take up the professorship in natural sciences at the newly established university in Berlin, now called the Humboldt University, although he did not have the required education (Lichtenstein was a physician). As a professor in Berlin he made special efforts to ensure a public role for natural history by establishing a museum of natural history in 1813 that was open to the public and by founding the Berlin zoo in 1844. He did not reach great heights as a scientist, but because of his South African journey and his account of this journey he was a popular social figure; for example, he dined with Wolfgang Goethe on several occasions and moved in the circles of the Royal Court in Berlin.[60]

A late postscript

A number of years after the end of the Dutch rule over the Cape, the "Dagverhaal van eene reis naar de Kaap de Goede Hoop, en in de binnenlanden van Afrika door Jonkvr. Augusta Uitenhage de Mist, in 1802 en 1803"[61] (Diary of a Journey to the Cape of Good Hope and the Interior of Africa by Lady Augusta Uitenhage de Mist, in 1802 and 1803) appeared in the Dutch ladies' magazine *Penélopé*. This diary was a fragment of a text that had been published anonymously in French for family and friends earlier and also covers Augusta de Mist's visit to America, the *Relation d'un voyage en Afrique et en Amérique par Madame* (1821).[62] The diary was not very newsworthy when it appeared, but the editor of Penélopé, Anna Barbara van Meerten-Schilperoort, neverthe-

[60] Mauersberger (1994).
[61] De Mist (1835, pp. 71–127). An English translation of this text was published (De Mist, 1954).
[62] According to the accompanying preface, a short run of this booklet was published for family and friends.

less considered it worth while for her readers because it "would not be unpleasant" and "to make some of my readers who occasionally venture to do a little travelling in summer more patient and content with the small mishaps that sometimes occur". The text was entertaining and instructive, which suited *Penélopé's* style.[63]

Although the diary of Augusta de Mist (fig. 39) was of minor interest as far as the information content was concerned, it is noteworthy from a gender perspective. It is the only text produced by a female author about a journey into the South African interior during this period. Today, especially due to Sarah Mills's *Discourses of Difference: An analysis of Women's Travel Writing and Colonialism* (1991) there is an expectation that women, because of their social position, had a different perspective of the colonial world and were more sympathetic to the indigenous population than Western men who were the bearers of colonial power. Although according to Mills women authors do not occupy a position outside the colonial discourse, in their texts there would at least be a conflict between this colonial discourse and the discourse of 'femininity', in which the latter would sometimes predominate.[64] However, in Augusta there is no trace of such conflict. Her views appear even more colonialist than those of male colleagues.

In her account, Augusta de Mist presents herself as the loving, still unmarried daughter who cannot allow her father to go travelling alone. She also generally regards arid landscapes as horrible, but they become "picturesque"[65] where there is more water and vegetation. Her ethnography largely follows the pattern set by other Cape travellers: the San are terribly ugly, cruel and like animals,[66] the Khoikhoi are lazy[67] and the Xhosas she also regards as admirable. For the latter view she relied

---

[63] For this magazine and its editor see: Gelderblom (1997, pp. 29–44). Below I use the Dutch text from the magazine *Penélopé*, because this is the version with the widest distribution and the most interesting from a gender perspective (see below). Augusta de Mist is mentioned as the author in this version (the French text was published anonymously as an independent publication); moreover, the Dutch text appeared in a ladies' magazine. Of the French text very few copies must have appeared, because via the Karlsruher Virtueller Katalog I was able to trace only one copy in the catalogue of the Bibliotheksservice Zentrum Baden-Würtenberg.

[64] Mills (1991, p. 63). Coetzee (1995, p. 69) has criticised Mills on the basis of the travel account by Augusta de Mist.

[65] De Mist (1835, p. 87).

[66] De Mist (1835, pp. 88, 97–98).

[67] De Mist (1835, p. 90).

completely on a book by captain Lodewyk Alberti, who will be discussed in the next chapter.[68] Unlike male travellers, however, she admires the colonists without reserve. She is pleased at their "patriarchal lifestyle, the goodness with which they treat their Khoikhoi and slaves, their diligence and their untiring labour".[69] Furthermore, the colonists "generally have pretty wives and not a few pretty children".[70] It sometimes seems as if she views the colonists from an early romantic, sentimental perspective, as in the dramatised description of the return of a husband:

> The next night we again spent under the hospitable roof of a colonist. He was absent, but his charming young wife, surrounded by five dear children, entertained us most excellently. We found in this accommodation great tidiness and the appearance of prosperity and domestic bliss. Nevertheless, our hostess could not completely conceal a certain feeling of anxiety. She was concerned about her husband, who, having gone hunting four weeks before, should have been back two days ago. She shared this distress with us, because she could see that we took an interest in her. Suddenly the door opens, and she lies in her husband's arms.[71]

Although Batavian travellers wrote about the colonists with more sympathy than travellers of another nationality, even they criticised abuses. In Augusta de Mist there is no trace of this at all. One explanation could be that her journey had no administrative or scholarly aim and it was therefore not important to determine the exact state of affairs in the interior. She lacked the empiricist urge that ran counter to current opinions. Consequently she lapses into a glorification, possibly inspired by the contemporary literature, of the simple and honest pastoral existence of the colonists. This approach had received a new impetus at the

[68] De Mist (1835, pp. 113–118). It is not entirely clear whether this is only an oral report by Alberti, whom De Mist met on her journey into the interior, or whether it is based on Alberti's *Kaffers aan de Zuidkust van Afrika* of 1810, a book De Mist could have consulted after her return. See chapter 9 for a discussion of Alberti's book.

[69] De Mist (1835, p. 87).

[70] De Mist (1835, p. 96).

[71] "Den volgenden nacht bragten wij weder onder het gastvrij dak van eenen Kolonist door. Hij was afwezig, doch zijne jonge bevallige vrouw, omringd door vijf lieve kinderen, schonk ons het uitmuntendst onthaal. Wij vonden in dit verblijf eene groote zindelijkheid en het voorkomen van welvaren en huisselijk geluk. Evenwel onze gastvrouw kon niet geheel een zeker gevoel van angst verbergen. Zij was ongerust over haren man, die, sedert vier weken op de jagt zijnde, reeds voor twee dagen terug had moeten zijn. Zij deelde ons dit verdriet mede, daar zij wel zag, dat wij belang in haar stelden. Daar opent zich eensklaps de deur, en zij ligt in de armen hares echtgenoots" (De Mist (1835, pp. 93–94).

end of the eighteenth century, partly through Rousseau, as a result of which Marie Antoinette had a small farm built in the Versailles palace gardens, for example.[72] Augusta de Mist created her pastoral idyll in the South African wilderness.

[72] For the glorification of the peasant's life at the end of the eighteenth century and the role of Jean-Jacques Rousseau in this, see Charlton (1984, pp. 178–198) and Van Tieghem (1960, pp. 213–221). Coetzee (1995, p. 70) argues—incorrectly, in my view—that Augusta de Mist followed the political project of the Batavian government of the Cape Colony.

CHAPTER NINE

THE FIRST ETHNOGRAPHIC MONOGRAPH: *DE KAFFERS AAN DE ZUIDKUST VAN AFRIKA* (1810) BY LODEWYK ALBERTI

In eighteenth-century travel accounts ethnography was a set part of a more comprehensive, encyclopaedic interest in the foreign world. People were described as well as landscapes, plants and animals. But between 1770 and 1790 the first signs of specialisation can be detected in the German-speaking countries and in Russia: ethnography started developing into a discipline.[1] Following on a comparable, somewhat later development in France, the first ethnographic monograph, *De Kaffers aan de Zuidkust van Afrika, Natuur- en Geschiedkundig beschreven* by Lodewyk (or Ludwig) Alberti (1768–1812) was published in the Netherlands.[2] Alberti's book offers a description of the western Xhosas which, at the time he wrote about them, occupied the country on both sides of the Fish River and part of the south coast of South Africa. Alberti's book was accompanied by a large picture atlas in oblong format, *Zuid-Afrikaansche Gezichten*, with three large ethnographic pictures out of a total of four.[3] In 1811 the French edition of both books was published. After Alberti's death in 1812 the original German text of *De Kaffers aan de Zuidkust van Afrika* was published as *Die Kaffern auf der Südküste von Afrika nach ihren Sitten und Gebräuchen aus eigener Ansicht bechrieben* on the initiative of his family in 1815.[4] The modern English edition is a free translation of this text.[5]

---

[1] According to current thinking cultural anthropology came into being only when anthropological museums (1836–1837) and ethnological societies (1839–1843) were established. However, Vermeulen (1995; 1996, pp. 5–16) has shown that specialisation started about seventy years earlier. For example, the terms *Ethnographie, Ethnologie, Völkerkunde* and *Volkskunde* were already used in Germany between 1770 and 1790.

[2] For an overview of the early Dutch ethnography see: Ellen (1976), Vermeulen & Kommers (2002). The Dutch edition of *De Kaffers aan de Zuidkust van Afrika* [...] was translated from the German by the professor of theology Konijnenburg (Alberti, 1810a, "Voorberigt").

[3] Alberti (1810b).

[4] Alberti (1815).

[5] Alberti (1968).

Alberti's book is unique in more than one respect. It is the oldest ethnographic description of the 'Kaffers', as the Xhosas were generally known at the time, the first Dutch ethnographic monograph and also the only practical application of the ethnographic questionnaire of Joseph-Marie Degérando (or De Gérando; 1772–1842), the *Considérations sur les diverses méthodes a suivre dans l'observation des peuples sauvages* (A consideration of the different methods to be followed in the observation of savage peoples) (1800). This questionnaire occupies an important position in the history of French anthropology as the earliest attempt to lay a theoretical basis for ethnographic descriptions. Lastly, *De Kaffers aan de Zuidkust van Afrika* is a very early example of applied anthropology in the sense of ethnography intended to serve administrative purposes within a colonial context.[6]

Although in South African historiography Alberti's book is an important source for the history of the Xhosas (for instance in the historical study of the Xhosas, Peires' *The House of Phalo*),[7] the book as such has never been researched. In this chapter I set out to explore the structure of *De Kaffers aan de Zuidkust van Afrika* and study the way in which Alberti followed the questionnaire and the ethnographic paradigm in Degérando's *Considérations*. I will use the Dutch version of Alberti's book as the basis of my investigation, as it is the oldest published version and the only one authorised by Alberti.[8] This version appeared at the time when Dutch claims to the Cape Colony were still valid. This is important to a third aspect I wish to investigate: the way in which Alberti attempted, by means of his book, to realise his ambition to be (re-)appointed as landdrost of the district of Uitenhage.

[6] Gulliver (1985, p. 37) defines the genre of applied anthropology as follows: "In my view [...] 'applied anthropology' involves both a particular orientation of research and a mode of presentation and use of its results. Research and its findings are specifically directed to the concerns of governments and their officers and are intended to be applied to policy making and administration." With regard to applied anthropology Kommers (1996, p. 97) refers to the "pretence" of an evident significance for colonial administration. Alberti indeed does not progress beyond pretence.

[7] Peires (1981). For his studies of the Xhosas of the Ciskei, Switzer (1993) did not even take the trouble to consult Alberti, but is satisfied with Peires' book (with thanks to Jean Kommers, who pointed this out to me).

[8] Alberti (1815, p. 202).

Alberti's career

Ludwig Alberti was born in the German county of Waldeck in 1768. He came to the Netherlands in 1784 as lieutenant of a regiment of Waldecker mercenaries and left for the Cape in 1802 as commander of a company of Waldeckers. In 1803 he and a small number of troops were posted far from Cape Town, at Fort Frederick, on the site of the present-day Port Elizabeth (figure 36). In October of that year Alberti became commander of the fort and in February 1804 'landdrost' (district administrator) of the newly proclaimed district of Uitenhage as well. Here he stayed until the Cape Colony was reoccupied by the English in 1806.

As landdrost of Uitenhage, Alberti had a particularly difficult task. He had to keep the peace in a volatile border district with a detachment that was far too small. Since the last years of the rule of the VOC the eastern border of the Cape Colony had been the site of confrontations between Xhosas, Khoisan and European colonists. In this conflict the Khoisan (the modern collective name for 'Hottentots' and 'Bushmen)' were the weakest party; by the end of the eighteenth century a large number of them had been assimilated, either voluntarily or by coercion, into the Xhosa community or worked for the European farmers as cowherds. The Xhosas and the European farmers expanded in opposite directions: the farmers eastward and the Xhosas westward. In the Zuurveld, an area between the Zondags River and the Fish River that had been regarded as the eastern border of the Cape Colony by the colonial authorities since 1778, these migrations collided, which gave rise to devastating border wars that continued far into the nineteenth century and eventually led to the subjugation of the Xhosas in the mid nineteenth century (fig. 4).[9]

The European farmers on the eastern frontier caused problems in other respects as well. Between 1795 and 1801 they rose up against the colonial authority three times. Despite the conflict with the Xhosas about grazing and cattle, the colonists sometimes concluded brief alliances with Xhosa tribes against the colonial government. After 1803 the Batavian government tried to stabilise this explosive border area. The main measure taken to this end was the proclamation of the new

[9] Peires (1981), Giliomee (1989), Mostert (1992).

border district of Uitenhage of which Alberti was appointed landdrost, although he had an insufficient number of troops at his disposal. Nevertheless, he carried out his task successfully. The main achievement of the short-lived Batavian government was that peace reigned on the eastern border between 1803 and 1806. For this Alberti deserves most of the credit. By regularly negotiating with the Xhosas and a clever divide-and-rule policy he was able to keep the warring parties apart.[10] After his departure in 1806 the border conflicts soon flared up again.

After the surrender of the Cape, Alberti returned to the Netherlands. He subsequently fought in Spain as a major in a Dutch detachment. After the publication of the Dutch edition of his book he accompanied the former Cape governor Janssens to Dutch India, where Napoleon—the Netherlands having become part of the French empire in July 1810—had appointed Janssens governor-general. In May 1811 Janssens arrived in Batavia with 500 men. His rule was short-lived, because on 3 August 1811 an English force landed on Java. After Batavia had fallen, Janssens retreated to the fort in Meester Cornelis. During the English attack on the fort (10 August) Alberti was hit by a bullet. A few months later, on 2 June 1812, Alberti died of 'fever'. He was buried in the cemetery of Weltevreden.[11]

Ethnographic system

Alberti wanted to realise several objectives with his book. In the first place the book is a letter of application. The book is dedicated to Louis Napoleon (1778–1846), Napoleon's brother and king of Holland (1806–1810) at the time the book was printed.[12] Alberti presents himself to the King in his "Voorberigt" (preface) as the future administrator of his old district of Uitenhage. Nowhere had he been happier in all his life than as landdrost of this tumultuous border district, he says. In 1810 such an open application was still possible, because the Cape Colony

[10] Freund (1971, pp. 319, 324).

[11] Alberti (1815, "Vorbericht"); Alberti (1968, pp. ix–xiv).

[12] The Kingdom of Holland ('Koninkrijk Holland') was created in 1806 by Napoleon Bonaparte as a vassal state of France, with Napoleon's third brother, Louis Napoleon Bonaparte, as head of state. Because Louis Napoleon did not live up to his brother's expectations—he tried to serve Dutch interests instead of his brother's—the Kingdom was dissolved in 1810, after which the Netherlands were annexed by France until 1813.

had been occupied only temporarily by England in 1806. The Cape Colony and other Dutch possessions officially became English territories only with the Treaty of London in 1814. In view of what Alberti says in the "Voorberigt", the book is intended to prove his competence as a district administrator. His competence consisted of ethnographic and administrative knowledge, with his knowledge as an ethnographer strengthening his capacity as an administrator. Nevertheless, the book is in the first place an ethnographic monograph.

The first twenty and the last two chapters of *De Kaffers aan de Zuidkust van Afrika* contain a description of the Xhosas. Most of this part is in the form of what is now known in anthropology as a 'realist tale', a description in which the authoritative author hides behind a list of facts about the daily life of the indigenous population.[13] Alberti emphasises the novelty of his information. According to him, the only publications on this subject that appeared before 1810 were Jacob van Reenen's report on a Dutch expedition to the Eastern Cape in search of survivors of the wreck of the English ship Grosvenor[14] and a passage in John Barrow's *Travels* of 1801.[15] In Alberti's view these texts were of no great value, because his predecessors had paid 'Kafferland' only brief visits and had not penetrated very far into it either. He himself, on the other hand, had travelled to 'Kafferland' repeatedly, and he had also noticed how his initial observations differed from those he made later.[16]

Alberti's list of predecessors is not entirely complete, because he fails to mention the description the Dutch missionary Johannes van der Kemp (1747–1811) had first published in the *Transactions* of the London Missionary Society in 1801. This description was reprinted in 1804 together with other reports by missionaries under the title "An Account of the religion, customs, population, government, language, history, and natural productions of Caffraria".[17] During his term as landdrost Alberti had had official contact with Van der Kemp.[18] He was probably not aware of the existence of this article, because I have been unable to identify borrowings from Van der Kemp's report. The ethnographic part of Van der Kemp's article was in any case a good

[13] Cf. Van Maanen (1988, pp. 46–54).

[14] Earliest publication in 1792 in the English translation by Riou; Van Reenen (1792).

[15] Barrow (1801, pp. 193–221).

[16] Alberti (1810a, "Voorberigt", p. ii).

[17] Van der Kemp (1804).

[18] Enklaar (1988, pp. 151, 155–156).

deal shorter than Alberti's book; it consists mainly of loose observations. But Van der Kemp did publish linguistic data,[19] a subject Alberti hardly deals with.

Alberti not only published the most comprehensive ethnography of the Xhosas of that time, but also went at it systematically. In his own words, he had attempted "to record that which everyone wanted to know about a nation that still seems to live almost in a natural state".[20] In practice he followed the guidelines drawn up by Baron Joseph-Marie de Gérando, or, during the period that is of interest here, 'citoyen' Degérando (1772–1842) in 1800 for ethnographic descriptions, the *Considérations sur les diverses méthodes a suivre dans l'observation des peuples sauvages*.[21] The most obvious connection between Degérando and Alberti is a quotation from the *Considérations* in the chapter in *De Kaffers aan de Zuidkust van Afrika* in which the question concerning the extent to which it is desirable to 'civilise' primitive peoples is discussed.[22]

When Degérando's *Considérations* and *De Kaffers aan de Zuidkust van Afrika* are put side by side, it becomes clear that Alberti relied on Degérando for the selection of his material as well as for the structure of his book. Like Degérando, Alberti confines himself almost exclusively to the study of the culture of the foreign people. Although he mentions a "physical" description in the title of his book, besides a "historical" description, the former does not amount to much more in practice than a short description of the area in which the Xhosas live (chapter 2), their external appearance (chapter 3) and their physical strength (chapter 5). The rest of the book is dedicated mainly to their culture. Here he follows Degérando's distinction between the description of the person as an individual and as part of the community. Specific issues that Degérando urged travellers to investigate turn up in Alberti—more or less in the same order as in Degérando. Table 2 illustrates these similarities.

Alberti is very dependent upon Degérando. When he formulates his aim as being "to record that which everyone wanted to know about

---

[19] Van der Kemp (1804, pp. 442–458).

[20] "[...] om dat gene op te teekenen, wat een ieder, eigenaardig, verlangt te weten van eene Natie, welke, tot hiertoe, bijkans nog in staat der natuur schijnt te leven" (Alberti (1810a, "Voorberigt", p. iii)).

[21] In the text below I always refer to the modern edition of the *Considérations sur les diverses méthodes a suivre dans l'observation des peuples sauvages* published by Copans and Jamin (De Gérando, 1978).

[22] Alberti (1810a, pp. 248–249).

a nation that still seems to live almost in a natural state",[23] then these are to a large extent the problems Degérando regarded as relevant in this regard in his *Considérations.*

Chapters from Alberti's *Kaffers* (1810a)		Degérando's *Considérations* (marginal notes) (1978)
Voorberigt		Premier défault (peu de séjour [...] au milieu d'eux) (133–124)
I	Origin of the Xhosas	Huitième défault (origine) (137), Traditions (166)
II	Location, extent and nature of the region of the Xhosas in the vicinity of the Cape	Climat (145)
III	Physical form and external appearance of the Xhosas	Etat des sauvages et d'abord de l'individu. Son existence physique (145)
IV	Food of the Xhosas	Aliments (145)
V	Physical strength of the Xhosas	Forces et actions physiques (146)
VI	Normal sleeping and resting of the Xhosas	Repos (146)
VII	Clothing and ornaments of the Xhosas	Vêtements (146–7)
VIII	Physical and moral education of children	Education physique (148), Education morale des enfants (157)
IX	Diseases, medicines and age of the Xhosas	Effets moraux de maladies (147) Longévité (148)
X	Language, drawing, writing and counting skills and mental capacity	Facultés (152–154)
XI	God, religion, superstition, withcraft and moral uncleanness of the Xhosas	Dieu; Esprits; Idées sur son existence; Immortalité (151–152)
XII	Domestic life and activities of the Xhosas	Société domestique (155)
XIII	Respect shown by children, family relations and respect for elders among the Xhosas	Autorité des pères (155)
XIV	Influence and morals of Xhosa women	Femmes. Leur état (155), Pudeur (156)
XV	Love and marriage among the Xhosas	Amour; Mariage; Son caractère, ses effets; Divorce, polygamie (156–157)
XVI	Social life of the Xhosas	

[23] Alberti (1810a, "Voorberigt", p. iii).

Table 2 (*cont.*)

	Chapters from Alberti's *Kaffers* (1810a)	Degérando's *Considérations* (marginal notes) (1978)
XVII	Government, respect, tributes and heredity of the Xhosa chiefs	Rapports politiques 1 intérieurs; Magistrats. Leur titre. Leur autorité (158)
XVIII	Jurisprudence of the Xhosas	Délits (161)
XIX	War and peace of the Xhosas	Guerre; Causes; Circonstances; Art militaire, Armes; Courage; Effets de la guerre, Paix, Alliance (159–160)
XX	Funerals and mourning among the Xhosas	Tombeaux (166)
XXI	Relations between the Xhosas and the colonists	–
XXII	Thoughts about the treatment of the Xhosas as regards the peace in and prosperity of the colony	–
XXIII	Thoughts about civilising the Xhosas	On pp. 248–249 Alberti quotes Degérando to counter the latter's opinions about civilising 'savages'
XXIV	Characteristics and particulars of Gaika	–
XXV	Popular view of the origin of the Xhosas	Huitième défault (origine) (137), Traditions (166)

Table 2: Parallels between Alberti's *Kaffers* [...] and Degérando's *Considérations* (1800)

Degérando and the Société des Observateurs de l'Homme

Degérando had drawn up his *Considérations* at the request of the Société des Observateurs de l'Homme to ensure that scientific travellers brought back ethnographic information that could be used for comparative anthropological research. This was prompted by the French expedition led by Nicolas Baudin (1754–1803) to Australia and a prospective new expedition by François le Vaillant to Africa. However, Le Vaillant never left on a new expedition, and the man who was responsible for anthropological observations during Baudin's expedition, François Péron (1775–1810), did not follow Degérando's guidelines.[24] In the literature on Degérando and the Société des Observateurs Alberti is not men-

[24] Stocking (1968, pp. 31–34).

tioned, although he was the only one who implemented Degérando's ideas, which occupy an important position within the history of French anthropology and the development of cultural anthropology.[25] Discussions of the implementation of Degérando's ideas to date have ended in a description of the failed application thereof by Péron. The successful implementation by Alberti has so far been overlooked.

Degérando and the Société des Observateurs de l'Homme play an important part in the history of anthropology, because they were the first to attempt to put the study of man and human culture on a theoretical footing. The Société was founded in 1799 to promote the scientific study of the "physical, moral and intellectual existence" of man. The activities of the society were short-lived, however, because it was already disbanded in 1804 as a result of internal problems. Its members included the naturalists Cuvier, Lamarck, Jussieu and Saint-Hilaire, the physicians Cabanis and Pinel and the travellers Bougainville and Le Vaillant. The young linguist Degérando was at that time one of the less well known members of the society.

The composition of the Society already shows that the approach to anthropology was encyclopaedic. No discipline enjoyed priority. Besides the physical characteristics, the cultural qualities of man were to be studied. With a view to the former, Georges Cuvier (1769–1832) drew up the guidelines for the collection of physical data on foreign people, the *Note instructive sur les recherches à faire relativement aux differences anatomiques des diverses races d'hommes* (Instructions for investigations to be conducted with regard to the anatomical differences between human races), for the Le Vaillant and Baudin expeditions. Here one could learn, for example, how to prepare skeletons and skulls (you had to boil them). Degérando gave guidelines for the collection of cultural data. The objective of the Observateurs was (useful) self-knowledge under the motto 'Gnoti te auton'—know yourself. By studying the Other, one could discover what all people have in common. The point of departure was a dualistic view of man. Following in the tracks of the naturalist Buffon and in opposition to the prevailing materialistic views of man (such as those

[25] Cf. Stocking (1968, pp. 13–41), Moore (1969), Moravia (1970), Copans & Jamin (1978), Chappey (2002). Chappey's comprehensive and thorough study of 2002, *La Société des Observateurs de l'Homme (1799–1804)*, is the most complete study of Degérando and the Observateurs, but once again no reference is made to Alberti. For the description of the ideas of the Observateurs I rely especially on Chappey.

of the Idéologues),[26] the Observateurs held on to the paradigm of the *homo duplex*, man as a combination of moral and physical characteristics, where the one cannot be reduced to the other. The Observateurs defended human dignity against attacks by the materialists. In the Observateurs' view, man's moral capacities were separate from the body. Subject to the primacy of reason, man was perfectible and not determined by physical characteristics. Scientific racism as it became current in the nineteenth century was precluded in this paradigm. The Observateurs held that human differences were not caused by the body but were of climatological, cultural, social or political origin. While they did draw an ethnocentric distinction between 'civilised' and 'savage' (on the whole Europeans were civilised, but some fellow Europeans could also be savage), the differences between these categories were gradual. All people formed one large family and shared one "Esprit". They differed from one another only in "intervals" of civilisation. In principle all of mankind went through the same development, some groups having merely advanced further on the long road of civilisation than others. This made the study of savages historically relevant. The savages were the contemporary ancestors of the more highly developed Europeans. One could see from the savages how one's own ancestors had lived.[27] That the savages had fallen behind the rest of humanity was, according to the Observateurs, a result of their geographic isolation. Besides studying savages, the Observers also saw it as their task to liberate savages from their isolation. The anthropology of the Observers was a philanthropic science with an explicit 'mission civilatrice'.[28]

The main outlines of the philosophy of the Observers were also present in Degérando's *Considérations*. The primary aim of his document, however, was to standardise anthropological research, to ensure that the scientific traveller—the *voyageur-philosophe*—took the necessary trouble and collected information that could be used for comparative anthropological research. Degérando wanted to construe the habitus of the observer.[29] His observer had to study the foreign society attentively,

[26] The Idéologues, such as Antoine Destutt de Tracy (1754–1836), author of *Elements d'idéologie*, viewed man's psychological faculties (the human 'ideology') as part of his physiology. According to Destut de Tracy, man's spiritual life could be traced back to sensory stimuli.

[27] This remained a persistent topos in subsequent 19th-century evolutionary anthropology. See Fabian (1983) in this regard.

[28] Chappey (2002).

[29] Chappey (2002, p. 268).

learn the language of his study objects and integrate himself into the foreign society as far as possible. Through prolonged residence he had to become a *concitoyen*, a fellow citizen[30]—an approach that brought Degérando close to modern ideas of ethnographic fieldwork. The study of man consisted, according to Degérando, of the three main parts, which are dealt with individually in the *Considérations*: the study of the physical nature and circumstances of the group observed (Degérando deals with this rather briefly), the study of man as an 'intellectual and moral being' in his different life phases[31] and (via a description of family relationships) as a member of the community of which he formed part.[32] These main categories are elaborated in the *Considérations* into more specific questions and subquestions. As part of the psychological characteristics of the individual, for example, resting/sleeping ("Repos") had to be studied. This gave rise to the following more specific questions:

> We must be told how many hours he [the savage] sleeps; whether his sleep is deep; whether he is at peace or troubled by dreams; what the nature of his dreams may be; whether he has a set hour for going to sleep; whether he finds being awake inconvenient or a nuisance; in what position he sleeps or rests.[33]

In the sixth chapter of his book, "Ordinary sleep and rest of the Xhosas", Alberti provides some direct answers to these questions.[34] Here we learn that the Xhosas are not fond of sleeping and that when they sleep, their sleep is peaceful and light; it is therefore easy to wake them up. In the morning they are well rested. "Otherwise these people very much like to be busy and enjoy almost no other rest than is necessary to satisfy their real needs."[35] According to Degérando's guidelines, such a description of the sleeping pattern would provide information about the mental characteristics of the foreign individual.

---

[30] De Gérando (1978, pp. 131–132, 135, 138).
[31] De Gérando (1978, pp. 138, 140).
[32] De Gérando (1978, p. 134).
[33] "On nous dira combien d'heures il accorde au repos; si son sommeil est profond; s'il est tranquille, ou s'il paraît occupé par des songes; quel peut être le caractère de ces songes; s'il a une heure fixe pour le sommeil; s'il supporte la veille sans déplaisir et sans incommodité; quelle est son attitude dans le sommeil ou le repos" (De Gérando, 1978, p. 146).
[34] Alberti (1810a, pp. 48–51).
[35] "Voor het overige zijn deze Menschen zeer gaarne bezig, en genieten bijkans geene andere rust, dan ter bevrediging hunner eigenlijke behoeften noodig is" (Alberti 1810a, p. 50).

Alberti's ethnographic innovations

In broad outlines, Alberti follows the main categories into which Degérando divided his topics. The physical living conditions and physical characteristics are discussed briefly in the beginning (chapters 2–5; fig. 40), after which the rest of Alberti's book describes the culture of the Xhosas, their moral and intellectual existence, in the words of Degérando and the Observateurs. Within this section, Alberti also follows the pattern set by Degérando by first examining man as an individual—sleeping pattern, clothes, education, in chapters 6–10)—and then proceeding, via their family relationships, to a description of the characteristics of their society (chapters 11–20; figs. 41–42).

Alberti is also aware of the quality requirements Degérando stipulated for a description. When he criticises Barrow for not having stayed in 'Kafferland' long enough, he knows that much depends on the frequency and duration of the contact. Alberti also notes that there is a huge difference between the impressions he gained after his first visit to the Xhosas and his subsequent opinions, after repeated visits.[36] This still falls far short of the requirement that Degérando set for the observer, namely that he had to become a 'concitoyen' of the people he observed, but it does show that Alberti was aware of this problem area. At the same time he thereby admits that he went through a learning process from being an outsider to becoming fairly familiar with the foreign culture, a topos also found in twentieth-century ethnography.[37] Furthermore, in the "Voorberigt" Alberti describes limitations under which he had collected his information. He did not speak the language of the Xhosas (Degérando attached great value to the mastery of language), so that he had to rely on interpreters.[38] Partly for this reason he was not able to determine much about the intellectual capacity of the Xhosas and the origin of their "superstitious acts".[39] Obtaining information on the

---

[36] Alberti (1810a, "Voorberigt", p. ii).

[37] Cf. Clifford (1988, p. 40) and Van Maanen (1988, pp. 75–78). The Nijmegen anthropologist Jean Kommers wrote to me as follows after my comments on Alberti: "In [twentieth-century] ethnography the learning process is often expressed in the compulsory distancing from touristic travelling: by staying longer or returning repeatedly you notice that your first impressions often lead to illusions: the tourist who 'knows' the other world after fourteen days. During fieldwork you recognize the illusions and you slowly start discovering 'what you do not know yet' [...]: exactly what Alberti experienced" (e-mail Jean Kommers, 14 April 2006).

[38] Alberti (1810a, pp. 86, 256).

[39] Alberti (1810a, "Voorberigt", pp. iii, iv).

latter was also impossible because they did not know anything about it themselves and because they had no priests, the group that normally would be able to provide information on matters of this kind.[40] Such an admission of one's own limitations also has a rhetorical function; it increases the credibility of other statements, which then have obviously passed the test of self-criticism.

Alberti had made use of indigenous informers. An important source was 'Gaika' or Ngqika (1775–1829), to whom he devotes a short chapter, "Characteristics and particulars of Gaika",[41] which intended to present a description of Ngqika's character by sketching situations and quoting statements showing Ngqika's prudence. At the time when Alberti was in contact with him, Ngqika was the most influential leader of the Western Xhosas. Earlier on Ngqika had been an important source of information for John Barrow.

In broad outlines, Alberti follows the scientific paradigm of Degérando and the Observateurs. Alberti is interested mainly in the culture of the foreign 'nation'; he sees this as a product of historical circumstances, which—as we saw earlier—he could not always identify.[42] In his view the Xhosas did not differ significantly from Europeans: they were not another 'race', let alone an inferior one, but rather a "wild or rather half-wild nation".[43] They were on a different civilisation interval, but formed part of the same humanity as the Europeans.

However, Alberti was sceptical of the *mission civilatrice* of the Observateurs. In one of the concluding chapters of his book he raises the question whether it is desirable to civilise them, as Degérando and the Observateurs would like to do. Alberti devotes the 23rd chapter: "Gedachten over de beschaving der Kaffers"[44] (Thoughts on civilising the Kaffirs) to this issue. Alberti reformulates this question as follows: "whether the true happiness of a wild, or rather half-wild nation is indeed promoted by civilising it or not".[45] The related question is whether the civilised or the uncivilised peoples should be regarded as happier. At first glance, "civilised man" is better off, but the savage is happier because he has fewer needs, which are therefore easier to

---

40 Alberti (1810a, "Voorberigt", pp. iv; 94).
41 Alberti (1810a, pp. 253–257).
42 Alberti (1810a, "Voorberigt", pp. ii, iii).
43 Alberti (1810a, p. 246).
44 Alberti (1810a, pp. 246–252).
45 "Of het waare geluk eener wilde, of liever half wilde Natie door hare beschaving inderdaad bevorderd worde of niet?" (Alberti, 1810a, p. 246).

meet. Civilisation creates new needs and reduces happiness. He then quotes (in translation) Degérando's exhortation to pass on the benefits of civilisation, but not the drawbacks:

> [Give them] our arts, but not our decay; our moral code, but not the examples of our vice; our sciences, but not our dogmas, about which we argue fiercely; the advantages of civilisation, but not its abuses: carefully hide from them how much people devour each other, even in the most enlightened countries, because of disputes and destroy themselves through their passions.[46]

On this Alberti comments as follows:

> If these instructions were faithfully carried out, the benefits of civilisation would be less subject to doubt: however, the extremes juxtaposed by Mr Degerando [*sic*] are so closely interrelated that it will not be easy to communicate the benefits without the disadvantages becoming intermingled with them at the same time.[47]

In other words: Alberti says that the advantages and disadvantages of civilisation are difficult to separate. It would therefore be cruel to change the happy and healthy situation of the Xhosas if in the process the disadvantages of civilisation are passed on.

At the end of this chapter Alberti makes this question even more complex by involving considerations of colonial administration. Assuming that civilising the Xhosas would be advantageous to them, it may only be done if it does not disadvantage the colony. In any case, missionaries must not be involved in this, because "such persons are usually profoundly ignorant, come from the lowest classes of the population and are religious fanatics throughout".[48] Alberti had a personal aversion to missionaries in any case.[49]

---

[46] "[Geef hun] onze kunsten, maar niet ons bederf; het wetboek onzer zeden, maar niet de voorbeelden onzer ondeugd; onze wetenschappen, maar niet onze leerstellingen, waarover men heviglijk twist; de voordeelen der beschavinge, maar geenszins hare misbruiken: verberg hun zorgvuldig, hoe zeer de menschen ons, zelfs in de meest verlichte landen, elkanderen door geschillen verscheuren, en zich zelven door hunne hartstogten vernielen" (Alberti, 1810a, p. 248); De Gérando, 1978, p. 132).

[47] "Bij eene getrouwe opvolging van dit voorschrift zouden de weldaden der beschavinge aan mindere twijfelingen zijn bloot gesteld: en echter staan de uitersten, door den Heer Degerando [*sic*] tegen elkanderen overgesteld, in zulk een naauw verband, dat de mededeeling der voordeelen niet ligtelijk zal kunnen plaats grijpen, zonder dat ook het nadeelige zich te gelijk daarmede vermenge" (Alberti, 1810a, p. 249).

[48] "[z]oodanige Persoonen zijn meestal diep onkundig, uit de laagste Volksklasse afkomstig en doorgaans godsdienstige dweepers" (Alberti, 1810a, p. 250).

[49] There was a personal reason for this dislike of missionaries. Alberti and the Batavian government were already irritated with the missionaries in the Cape Colony in

With this parting thought Alberti takes his leave of Degérando's themes. Colonial administration plays no part anywhere in Degérando because he focuses on the description of wild people outside a colonial context. Alberti, on the other hand, deals fairly extensively (37 pages in total) with the troubled relations between the Xhosas and the Cape Colony in chapters 21 and 22: "Betrekkingen tussen de Kaffers en de Volkplanting" (Relations between the Kaffirs and the Colony)[50] and "Gedachten over de behandeling der Kaffers ten opzigte van de rust en welvaart der Volkplanting" (Thoughts about the treatment of the Kaffirs in regard to peace and prosperity of the Colony).[51] The former chapter contains a historical overview of the border conflict up to the departure of the Batavian administrators in 1806. The next chapter outlines a policy for the border region. In these chapters Alberti is no longer the neutral ethnographer, but assumes the persona of an experienced colonial administrator with relevant ethnographic and political knowledge.

According to Alberti's historical overview, the conflict on the eastern border of the Cape Colony is not due to the Xhosas.[52] He follows the premise that the Xhosas as a savage people behave well when they are treated well. Unfortunately the "Christian neighbours" in the colony have not treated the Xhosas so well, which has had a corrupting effect on them. He had already expressed this opinion in a letter to governor Janssens in 1805.[53] In his book Alberti wrote that the problems were aggravated by divisions among the Xhosas, an anarchist attitude among the 'Jacobite' colonists towards the government, an alliance between the Xhosas and the Khoikhoi who had been maltreated by the colonists and an unbalanced policy of the temporary English government

1804, especially with Van der Kemp, because he advised the Khoikhoi not to become soldiers (the Batavian government had a regiment of indigenous troops) and was reluctant to make the Khoikhoi who were in his care available as labour for Fort Frederick, of which Alberti was the commander (cf. Freund, 1971, pp. 381, 384; Enklaar, 1988, pp. 151, 155–156, 161). Alberti had also in 1804 advised against allowing Van der Kemp to do missionary work among the Xhosas (Freund, 1971, p. 391).

[50] Alberti (1810a, pp. 207–233).

[51] Alberti (1810a, pp. 234–245).

[52] Alberti (1810a, pp. 207–233).

[53] "Nobody can be more convinced of the incompetence of Europeans to deprive peaceful peoples of other parts of the world of the possession of their original places of residence in order to gain some political or financial advantage from it than I am; this conviction cannot result in anything else than utter disgust at all the unnecessary murders or even only maltreatment of individuals or nations living in this outpost" (letter from Alberti to Janssens, 12 June 1805, CA, BR 377, f. 67).

(1795–1803). It was only with English force (before 1803) and much Batavian diplomacy (1803–1806) that it had been possible to temporarily restore the peace. For a permanent arrangement, however, it would be necessary to evict the Xhosas who had settled within the borders of the colony. Owing to the threat of war, the Batavian government was unfortunately unable to do this.

In the next chapter, "Gedachten over de behandeling der Kaffers ten opzigte van de rust en welvaart der Volkplanting",[54] Alberti provides a blueprint for a permanent solution to the problems in the border area. These proposals contained two key points: expelling the Xhosas from the colony and delegation of authority to the landdrost of the border district of Uitenhage, the position Alberti had occupied and again aspired to. Once the Xhosas had been driven out, it would be the landdrost's duty to keep the warring parties apart and to maintain good relations with the Xhosas. "This landdrost should be an energetic person, diligent in serving the Colony, but also of an enlightened mind about uncivilised nations and fair in his treatment of them."[55] He subsequently explains this "enlightened mind about uncivilised nations". 'Enlightened' meant that the landdrost of Uitenhage should possess ethnographic knowledge. He had to know how Xhosas think and thus be able to press the right buttons in order to achieve the desired effect. The example he uses to illustrate his thought is typical. He wanted this example to show how ethnographic knowledge helps to exercise power:

> Lastly, like knowledge of human nature in general provides the greatest certainty about all acts between one person and another, one should also know the mind of a wild or semi-wild nation well in order to influence it to advantage and to achieve the desired goal. It is with regard to this that the preceding considerations produce adequate conclusions to serve as the basis of subsequent acts. For example, the Kaffers hold the will of the supreme power in high esteem, which they are accustomed to respect. Very different, therefore, is the impression which the prohibition of the governor of the colony makes on them from that of the colonists when the latter put some request or other to them, and for this purpose use an imperative tone of voice.[56]

---

[54] Alberti (1810a, pp. 234–245).

[55] "Deze landdrost behoort een werkzaam man te zijn, ijverig voor den dienst der Volkplanting, allezins verlicht van denkwijze omtrent onbeschaafde Natiën en billijk in derzelver behandeling" (Alberti, 1810a, p. 242).

[56] "Eindelijk, gelijk menschenkennis in het algemeen de zekerste regelmaat verschaft omtrent alle handelingen van mensch tot mensch, ook alzoo behoort men den geest

In other words: like "knowledge of human nature"[57] is indispensable in bourgeois society, you must also know the "mind of a wild or semi-wild nation" in order to follow an effective policy with regard to such a nation. For example, if you know to whom the Xhosas have assigned authority, you also know as colonial administrator how you can exercise authority over them. Ethnographic knowledge is therefore indispensable to an effective administrator, and this knowledge is contained in Alberti's book. It is also striking in this pronouncement that Alberti does not in principle draw a distinction between ethnographic knowledge and bourgeois *savoir vivre*. Both are means of self-assertion; the European world is therefore not materially different from that of the "wild nations".

Conclusion

Alberti adroitly brought the threads together in the concluding chapters and proved his eligibility for the office of landdrost of Uitenhage. The first twenty chapters testified especially to his ethnographic knowledge, after which he stated in the 22nd chapter that this knowledge was indispensable to someone who had to perform the duties of the landdrost of Uitenhage. In the conclusion he presents himself as a pragmatic administrator to whom the interests of the colony come first. No matter how pure the character of the Xhosas as a "wild or semi-wild nation" may be, the contact with the hypocritical colonists has unfortunately spoilt them, and it is better to push them beyond the border. According to Alberti, they even penetrated far into the colony on their hunting

ener wilde of half wilde Natie naauwkeuriglijk te kennen, om op derzelve met voordeel te werken en het bepaalde oogmerk te bereiken. Het is ten dezen opzigte, dat de hier voorgaande beschouwingen genoegzame gevolgtrekkingen opleveren, om tot grondslag van volgende handelingen te dienen. Dus, bij voorbeeld, hebben de Kaffers een zeer hoog denkbeeld van den wil der Oppermagt, welke zij gewoon zijn te eerbiedigen. Zeer verschillende is daarom de indruk, dien het verbod van den Gouverneur der Volkplanting op hen maakt, van dien der planters, wanneer deze het een of ander ernstig van hun verlangen, en daartoe zelfs een bevelenden toon bezigen" (Alberti, 1810a, p. 244).

[57] The *Woordenboek der Nederlandsche Taal op CD-Rom* defines 'menschenkennis' (knowledge of human nature) as follows: "Duidelijk inzicht in het karakter der menschen in 't algemeen, of in dat van de menschen binnen een bepaalden kring" (having a clear understanding of the nature of people in general, or the nature of people in a specific social circle).

trips, which caused unrest among the colonists living there. The paradoxical position of the colonial official as ethnographer already comes to the fore in this early ethnographic description. Although Alberti is sympathetic to the Xhosas and their leader Ngqika, in the end the *raison d'état* wins.

Alberti's book was no mean feat. Somewhere between 1800, when Degérando's *Considérations* appeared in Paris, and before 1810, the year in which *De Kaffers aan de Zuidkust van Afrika* was published, he came into possession of the latest and at that time the best ethnographic questionnaire that provided him with a guideline for the selection of his material and the structure of this book. Alberti was the only one who successfully applied this questionnaire. However, it is he who deserves the credit for realising—based on his practical experience—that ethnographic knowledge can be relevant to colonial administration. This made his work not only the first theoretically founded ethnography to be published in the Netherlands, but at the same time it gave an initial impetus to applied anthropology. It was no more than an impetus, because the application of ethnographic knowledge remains limited to one example that must in the first place serve to illustrate Alberti's competence as an enlightened colonial administrator. Alberti used it to show how suitable he was to be appointed landdrost of Uitenhage once again, so that he could return to the region where he had spent the happiest time of his life: "To be allowed to return to the southern tip of Africa, so as to continue my service and my research, would certainly be the fulfilment of one of my most fervent wishes."[58]

[58] "Naar Afrika's Zuidpunt, ter voortzettinge van mijnen dienst en van mijn onderzoek, te mogen wederkeeren zoude gewisselijk de vervulling van eenen mijner vurigste wenschen uitmaken" (Alberti, 1810a, "Voorberigt", p. viii).

CHAPTER TEN

CONCLUSION: KNOWLEDGE AND COLONIALISM

Introduction

"Where, one asks, is everybody? The landscape is written as uninhabited, unpossessed, unhistoricized, unoccupied even by the travelers themselves."[1] According to the author of this statement, Mary Louise Pratt, scientific travellers were in the habit of omitting indigenous people and even themselves from their descriptions. They would describe panoramic views of deserted landscapes in preparation for colonial expansion and capitalist exploitation. When they did notice people, they would have been taught by Linnaeus and Buffon to represent them as natural phenomena. They also kept silent about their own presence in order to strengthen the impression of objectivity. In their books they are nothing more than a travelling eye that records nature. Only Peter Kolb had taken the full humanity of the indigenous population into account because he lived at a time before the natural history of Linnaeus and Buffon—the cause of this assumed reductionism—was able to exert its pernicious influence. Because of this rhetorical form of 'ethnic cleansing', Pratt blames eighteenth-century science as co-responsible for colonial cruelty in South Africa.[2] But as we saw—especially in the chapter on John Barrow—Pratt presents a loaded picture of the representations produced by scientific travellers, and moreover she overestimates the political effect these representations could have had.

Contrary to Pratt's assertions, ethnography was to a limited degree even part of the interest in the natural history of the foreign world. For example, in the section "Oeconomica" of Linneaus' 'Manual for

---

[1] Pratt (1992, p. 51).

[2] In English language studies of travel accounts this view has many supporters. Pratt's view of scientific travellers (Pratt, 1992, pp. 1–68) is supported among others by Musgrove (1999, p. 36), Wheeler (1999, p. 19), Blanton (2002, p. 12) and Fulford, Lee & Kitson (2004, pp. 12–13). The electronic database Web of Science contains almost a thousand references to Pratt's *Imperial Eyes* (April 2006), which is an indication of the influence of her work. See also the following critiques of Pratt: Beinart (1998); Huigen (1998); Guelke & Guelke (2004).

the [scientific] traveller', the *Instructio Perigrinatoris* of 1759, he actually urged travellers to investigate the physical living circumstances of the inhabitants of an area.[3] Ethnography is also covered in Robert Boyle's natural history questionnaire, the "General Heads for the natural history of a Countrey, Great or small", first published in 1666 in the *Transactions* of the Royal Society. Among other things, Boyle says: "[...] there must be careful account given of the *Inhabitants* themselves [...]", and their "*Traditions*" had to be taken into account at the same time.[4]

If the indigenous population had been omitted from the representation in the texts analysed by Pratt or if they had been reduced to natural phenomena (which is not the case), it still remains very much a question whether this would have encouraged colonial expansion, as Pratt assumes. The colonial process of dispossession in South Africa had been in full swing for more than a century by the time the texts analysed by Pratt were published and was driven by socio-economic motives—especially by the colonists' constant need of land for extensive stock farming.[5] The reading matter of the trekboere (migrant farmers), the group mainly responsible for this colonial expansion, was at best confined to the Bible or to Calvinist tracts and did not include travel accounts.[6] Colonial office bearers, such as Abraham de Mist, read travel accounts in order to increase their knowledge of the area they had to govern, but these reports did not in the least encourage them (as Pratt suggests) to expand the colony.[7] They merely wanted to know more about the land that was already considered part of the colony. Reading the travel accounts actually made De Mist critical of the colonists, the very group that had an interest in further expansion.

In eighteenth-century travel accounts on South Africa there is in any case no question of remaining silent about the indigenous population. As appears from the preceding chapters, most travellers in South Africa combined their interest in nature with ethnography in a broader encyclopaedic and not strictly naturalist research programme. Foreign

[3] I used the edition of the *Instructio Perigrinatoris* in Linnaeus (1760). For comments on the *Instructio* see: Kury (2001, pp. 109–110).

[4] Boyle (1666, pp. 188–189); original italics.

[5] The colonial expansion has been adequately explained in the historical literature. A reference to the essays in *The Shaping of South African Society, 1652–1840* (Elphick & Giliomee, 1989) will suffice.

[6] Van Zijl (sa).

[7] Pratt does not discuss the reception of travel literature. She merely makes allegations about the rhetorical effects of travel accounts.

people received their full share of attention in this research programme. In Alberti's work, ethnography is even the sole subject. Scientifically educated European travellers were actually the first to take an interest in the indigenous peoples of South Africa. The seventeenth-century passers-by who called at the Cape long 'before Linnaeus' stuck to stereotype descriptions of the semihuman Hottentots; only in 1719 did Peter Kolb turn them into humans. Especially researchers of the last quarter of the eighteenth century—Gordon, Le Vaillant and Barrow—and the first years of the nineteenth—the Batavian travellers, especially Alberti—followed in his tracks and extended anthropological interest to the Gonaquas, Xhosas and Tswanas. As part of their intention to describe *everything*, representations of South Africa also devoted attention to the culture of the indigenous population. The degree to which this was done with respect for the otherness of indigenous people differed, however. It was generally found difficult to view the San as completely human. Kolb and Le Vaillant restored the status of the Khoikhoi up to a point, but because the latter sank into a hopeless state of servitude in the course of the eighteenth century their fate was lamented with a shiver of repugnance by most travellers. Le Vaillant was able to make the Khoikhoi 'salon-ready' only by adapting them to European ideals of beauty and turning them into objects of primitivist wishful thinking. Only the Xhosas and Tswanas in the eastern and northern Cape were almost without reserve experienced as admirable, even if writers were unable to refrain completely from adapting them to European tastes.

The efforts to correct the image of the indigenous population positively had effects even beyond the circles of those who occupied themselves specifically with South Africa. There is a connection, not often mentioned, between Cape ethnography and the philosophical vanguard of the Enlightenment in the persons of Diderot, Kant and Herder, philosophers who have recently been pointed out as "anti-imperialist, political thinkers" because of their critical attitude to colonialism and their ability to view foreign cultures as equivalent.[8] Via direct reception of Kolb's works, as by Kant,[9] or indirectly via everyone in the enlightened Europe who had read Jean-Jacques Rousseau, Kolb's ethnography influenced the image of primitive societies that these philosophers had made up for themselves. Rousseau, in his turn, again played a part in

[8] Muthu (2003).
[9] Firla (1994).

the representation of indigenous people in southern Africa at the end of the eighteenth century. In his *Voyage* (1790), Le Vaillant sculpted his image of the admirable Hottentot after the example of Rousseau's noble savage.

Diderot, Kant and Herder arrived at the conclusion—radical for its time—that cultures are not mutually comparable; one culture could therefore not be measured by the standards of the other.[10] Travellers in the Cape found this idea harder to swallow. An attitude of unqualified tolerance towards other forms of existence in eighteenth-century South Africa can only be found in Robert Jacob Gordon, the most empiricist of the scientific travellers. There is no condescending remark about indigenous people in his works. Gordon viewed even the San, detested by other travellers, as fully-fledged humans with an intriguing culture of their own. Partly from a naturalist perspective, he concluded that people generally resemble each other more than animals that according to the eighteenth-century science belonged to the same species. Contrary to Pratt's contention, the naturalist perspective could therefore also lead to an emphasis on structural correspondences between people who at first glance differ greatly from each other.

Although Pratt is mistaken when she presents scientific travellers as trailblazers for colonial expansion in South Africa, they and other travellers were most certainly dependent on cooperation with the colonial regime. They owed their knowledge of South Africa and its residents partly to the colonial regime and to the colonists they often found repugnant. At the least, colonial expansion made the interior of South Africa more accessible to scientific travellers. The relations between colonialism and knowledge are even closer when scrutinised more carefully. In order to understand the interconnectedness of colonialism and the gathering of knowledge in the eighteenth century in South Africa better, however, we need to change our analytical perspective and focus our attention on the networks within which the representations play a part. The extent of the collaboration only becomes clear when one examines the way representations—including representations which were critical of the colonial dispensation—were produced within cooperative ventures.

Scientific travellers could carry out their activities only by obtaining support from colonial bodies and individuals. In different circumstances,

---

[10] Muthu (2003, pp. 259–283).

this is not much different for modern-day scientists. According to the French anthropologist of science Bruno Latour, scientists' skill in obtaining support outside the domain of science is an important key to their success.[11] For a better understanding of the interrelatedness of colonialism and the production of knowledge we must therefore investigate how the interaction between the travellers and the colonial dispensation worked. For this I will refer to all types of travellers who have been discussed in this book.[12]

Below I will distinguish several Cape exploration networks, i.e. groups that cooperated to collect information on the interior of southern Africa.[13] Each network went after a different kind of knowledge: the commercial network of the VOC as a trading company (1652–1795) aimed to collect commercially relevant information; the administrative network wanted to collect data relevant to the administration of the region; the scientific networks occupied themselves with knowledge that was regarded as relevant within the European scientific community at that time. In the case of the VOC the first two networks overlap. The VOC was a commercial enterprise, but it also acted as a colonial state. Facts collected by one network could also be important to another network. For example, the giraffe was of potential interest to the VOC as a beast of burden, but it was also (and especially) an important discovery for natural history. After a concise description of the Cape exploration networks, I will also pay attention to the epistemology that was used within the networks, especially the scientific networks.

---

[11] Bruno Latour says the following with regard to Louis Pasteur: "He who is able to translate others' interest into his own language carries the day. [...] So it is useless to look for the profit that people can reap from being interested in Pasteur's laboratory. Their interests are a consequence and not a cause of Pasteur's laboratory. Their interests are a consequence and not a cause of Pasteur's efforts to translate what they want or what he makes of them" (Latour, 1983, p. 144).

[12] The analytical perspective I follow is derived from the actor network theory (ANT for short) that was developed since the eighties by Michel Callon, Bruno Latour, John Law and others to investigate the activities of scientists: Latour (1983); Latour (1987); Latour (1990); Latour (1996); Latour & Woolgar (1986); Callon (1986); Law (1986).

[13] The application of the actor network theory to the activities of explorers is not new. Two of the three most prominent ANT researchers have already occupied themselves with it. Bruno Latour (1987, pp. 215–232) described the mapping of the island of Sakhalin by the eighteenth-century French explorer La Pérouse and John Law (1986) investigated the ships of fifteenth- and sixteenth-century Portuguese explorers, with their crew, equipment and navigating abilities, as networks.

The commercial exploration network

The oldest exploration network was that of the VOC. From 1660 until the end of its rule in 1795 the VOC investigated the economic potential of the interior of South Africa. In order to collect reliable information an exploration network was developed as early as the seventeenth century. This network was given a definite shape during the expedition that left for Namaqualand and was headed by Commander Simon van der Stel in 1685–1686. Subsequent expeditions followed this model and in some cases were even given Van der Stel's account of his journey as a reporting template. The commercial exploration network was characterised by a hierarchical structure in which those who journeyed into the interior were expected to be extensions of the principals, who in this way hoped to acquire reliable information.

The procedure was as follows. After an instruction or approval had been issued by the Gentlemen XVII in Amsterdam (Heren XVII, the directors of the VOC) or by an authorised commissioner, the Council of Policy, the local administration at the Cape, appointed the employees who would carry out the expeditions and drew up a "memorandum" or "instruction" that specified how the expedition was to be conducted and what was to be investigated. The memorandum had to ensure that the expeditions were transparent to the principals, that the participants in expeditions were as far as possible executors of the will of their principals. The travellers' freedom of action was limited by the institutional framework. The purpose of such an expedition was usually to explore specific commercial possibilities (trading gold with Monomotapa, mining copper in Namaqualand). In each case the expedition had to bring back a report, which comprised mainly a route description, and preferably also a map of the route followed.[14] The distinction between text and map was not precise, as the travel journal had a strongly cartographic character because of the regular indication of distances covered and the direction taken.[15] On the basis of such a VOC route description, for example, the English captain Riou was able to produce a map of the

[14] The commercially driven exploration network of the VOC in the seventeenth century is described more comprehensively in Huigen (1996a, pp. 23–59).

[15] Most of the VOC travel journals can be called cartographic texts, because they serve the same purpose as route maps. For the relation between route description and map, see: Hoogvliet (1999).

route followed by an expedition in 1790–1791 to the place where the English ship, the Grosvenor, had run aground.[16]

The Van der Stel model furthermore determined that VOC expeditions had to be large. The expeditions consisted of a great number of troops supplied with wagons and draft or pack-oxen. This restricted the freedom of movement of the expeditions considerably, because provisions taken along were eventually exhausted or because there often was not enough water for a long caravan. The only advantage was that the expedition would generally outnumber any indigenous opponents.[17] Before Hendrik Hop's overland journey ('landtocht') of 1761–1762 the expeditions generally consisted of soldiers. After 1761 the troops were mainly colonists, which reduced the costs of expeditions. Nevertheless, the output of the VOC overland journeys was small in all cases. At best the long caravans brought back reliable route descriptions, and in many cases a map as well. Occasionally travel journals of VOC overland journeys were published outside the commercial VOC network. Thus they could become part of scientific networks.

It remains remarkable that a trading company such as the VOC stuck to this expensive form of exploration until the end of the eighteenth century. In the meantime, individual scientific travellers had booked much better results with small-scale and relatively autarchic forms of organisation. Scientific travellers usually needed only one or two wagons, a European servant and a few Khoikhoi. This gave them a higher mobility and a greater range, as they were able to sustain themselves by hunting. The information they brought back in the form of diaries, specimens, drawings and cartographic surveys ("immutable mobiles" in Latour's terminology)[18] were also of a higher quality and greater scope than the cumbersome VOC expeditions were able to produce.

In conclusion we may say that the commercial exploration network of the VOC was characterised by very limited objectives, strict organisation and low yield. As a trading company, the VOC needed only limited information on the interior. Scientifically relevant knowledge was looked for only occasionally.[19] This was not to the satisfaction of everyone at

---

[16] Riou added the map he had drawn to the travel journal of Jacob van Reenen, which he had translated and published (Van Reenen, 1792).

[17] As we have seen, this did not apply to the expeditions that travelled into the interior from Fort Lijdsaamheijd.

[18] Latour (1987, p. 227).

[19] In South Africa, for example, by order of Hendrik Adriaan van Reede tot Drakenstein en Nicolaas Witsen (see: Huigen, 1996a, pp. 31–33, 54–55). For the

the VOC. The VOC director Nicolaas Witsen (1641–1717), who took a strong interest in science, complained to a friend in 1717: "Why does your Honour ask about scholarly curiosity in India? No, Sir, it is only money and no knowledge which our people are after, which is regrettable."[20]

The administrative exploration network

Besides being a trading company, the VOC also acted as a colonial state towards indigenous communities and within the area it claimed as its own. For its administrative tasks it needed information on the constantly growing Cape Colony, which was governed from Cape Town and regional 'drostdijen' (magistrate's offices). The successors of the VOC, the English and Batavian governments, needed even more information, especially about the eastern border area, which had been in a state of almost permanent turmoil since 1779. The Batavian colonial government had an additional need for information because it saw itself as an administration that wanted to promote the well-being of "fellow citizens", the colonists. This could be achieved effectively only if the administrators were familiar with the living conditions of their citizens. Two important means of getting an administrative grip on remote areas of the colony were inspection journeys by high-level officials and the production of topographic maps.

The VOC governor Van Plettenberg, commissioner-general De Mist and the Batavian governor Janssens hoped to give their policy an empirical basis by criss-crossing a large part of the colony, holding discussions with inhabitants and visiting the problematic border area, where negotiations were conducted with the Xhosas. In undertaking these inspection journeys and developing a policy the colonial authority relied partly on the travel accounts of scientific travellers. This shows how texts of this nature could be transferred from the scientific network to the administrative network. This transfer took place because

involvement (often reluctant) of the VOC in scientific projects, see: Van Berkel (1998, pp. 131–146).

[20] "Wat vraegt UwelEd. na de geleerde curieusheyt van Indiën, Neen Heer, het is alleen gelt en geen wetenschap die onse luyden soeken aldaer, 't gunt is te beklagen" (Gebhard, 1881–1882, II, pp. 340–341 (letter 1 August 1712; also quoted in: Van Berkel, 1998, p. 145). Witsen was known among other things as the author of *Noord en Oost Tartaryen*, (1705) in which he described Siberia.

the scientific travel accounts contained information that was useful for administrative purposes, not because of rhetoric that encouraged colonial expansion, as Pratt claims.[21] De Mist, for example, had drafted his colonial policy for the Cape Colony in the Netherlands on the basis of VOC archives and the first part of John Barrow's *Travels*, although afterwards he was annoyed at the criticism Barrow levelled at the Dutch colonial dispensation. On their journey into the interior Lichtenstein and De Mist even had a fairly complete library of travel writings at their disposal. The information they collected was used mainly to correct and supplement the existing representations. It is also striking that the travel journals of officials such as De Mist and Paravicini di Capelli have much in common with scientific travel accounts as regards their wealth of information and their broad perspective. By the end of the eighteenth century scientific travellers had shaped the habitus of the traveller and the form of the travel account to such an extent that administrative officials adopted it.

The administrative network offered work to people with an interest in science, who in many cases made use of this network for their own purposes.[22] Gordon served as an officer in the VOC, Barrow was secretary to the English governor Macartney at the time he made his most important South African journeys, Lichtenstein was for the greater part of his stay at the Cape officer of health in the Batavian army and Alberti was an army captain and acting landdrost of Uitenhage during the same period. The writings of these four researchers were of little political or administrative value. Gordon's works and Lichtenstein's publications were mainly of a scientific nature. Only a small part of Alberti's *De Kaffers aan de Zuidkust van Afrika* shows how anthropological knowledge can be used for administrative purposes; the major part of the book was conceived with the aid of Degérando's *Considérations*, which aimed to instruct travellers to collect scientifically useful ethnographic

[21] The point is that Pratt does her utmost to prove the bad faith of scientific travellers. She says that their actions were aimed at encouraging colonial expansion while they hid behind a screen of innocence, an attitude that Pratt calls "anti-conquest". That colonial administrators such as De Mist made use of travel accounts proves nothing about the intention with which these travel accounts were written. Actually Pratt does not even mention this use of travel accounts by the colonial government. She deems it sufficient to assume that evil intentions are found in the travel accounts written by scientific travellers.

[22] For many, scientific curiosity was a passion. According to Hobbes scientific curiosity was even a desire and as much a characteristic of man as his intellect (Daston & Park, 2001, p. 307). I will return to this later.

information and did not take account of a colonial situation. The official end of Dutch rule over the Cape Colony in 1814 deprived *De Kaffers aan de Zuidkust van Afrika* of any political significance. Lichtenstein integrated fragments from the travel journals of commissioner-general De Mist and governor Janssens, which had been written with administrative purposes in mind, into his *Reisen*; in this way these originally official reports acquired scientific significance.

The knowledge gathered by the researchers mentioned above was related to existing scientific problems and acquired a place within European science. The impact of their knowledge was greater in Europe than within the small administrative network in the Cape Colony, which was in a state of constant flux in the period 1795–1814. Only the second part of Barrow's *Travels* of 1804 had a political objective. With this book, Barrow wanted to persuade English politicians and public opinion that the Cape should again be put under English rule by proving the strategic military and economic value of the Cape.

An important aspect of administrative activity during the last decades of the eighteenth century was the compilation of maps. These were not route maps of reconnaissance expeditions that put only a narrow strip of land on both sides of the route on the map, like the expeditions the VOC sent into the interior did, but topographic maps drawn with the intention of providing a complete overview of the Cape Colony. There is a direct connection between the production of these topographic maps in South Africa at the end of the eighteenth century and the endeavour to define the borders of the Cape Colony.[23] The latter was in practice a serious problem in the east of the colony, where colonists and Xhosas got in one another's way and the colonial government wished to establish a clear demarcation in order to bring the never-ending conflicts to an end. This effort met with little success, so that until deep into the nineteenth century bloody border wars were fought. The problem was that it proved very difficult to make border agreements with the Xhosas, because the Xhosas had no central authority and the power of the colonial government on the eastern border was very limited up to the time of English rule. The complicated geography there (mountains and dense forests), probably in addition to interpreting problems during negotiations with the Xhosas, made it difficult to establish the border in 1778. In South African historiography, it is

[23] See also Penn (1993).

generally assumed that part of the Great Fish River constituted the border between the Cape Colony and 'Kafferland' since 1778, but even the VOC governor Joachim van Plettenberg seems not to have been quite sure which border he had agreed in 1778.[24] In a personal letter to his friend Hendrik Swellengrebel dated March 1780 he seems to suggest that his border agreements would not stand up juridically and would not convince the Gentlemen XVII either.[25] However, he hoped that the new topographic maps made by the surveyor Leiste would clear up the doubts about the border agreements (fig. 43):

> [...] the surveyor Leijste is actually busy with the map, which I intend to present to my masters [i.e. the Gentlemen XVII] in the expectation that via this channel [the map] will be made public, which I have no liberty to do. The main bays I will add on a large scale for the information of shipping, which is quite unfamiliar with our coast.[26]

The surveyor C.H. Leiste produced several topographic maps of the Cape Colony, but on his maps too a clear indication of the eastern border is absent.[27] The first topographic map that showed the entire territory of the colony and that could serve as a basis for administrative and defensive measures was made on the instruction of Van

[24] A discussion of the border problems can be found in Bergh & Visagie (1985).

[25] Van Plettenberg writes to Swellengrebel that he doubts that his border agreements will stand up to criticism: "En selfs dunkt mij, dat deeze quaestie [de grens] eens in een Disputeer Collegie wordende voorgesteld, een welspreekend opponens aan den Seer geleerden Heer Defendens nogal werk soude verschaffen" (Schutte, 1982, p. 109). In a resolution of the Council of Policy of 14 November 1780 it is stated that it had been agreed with the Xhosa chiefs that the Fish River formed the eastern border of the Cape Colony (Jeffreys, 1928, pp. 93–94).

[26] "[...] de Landmeeter Leijste is werkelijk besig met de Kaart, die voornemens ben mijne Heeren Meesteren [i.e. de Heren Zeventien] aan te bieden, in verwagting dat door dat Canaal [de kaart] publijcq sal gemaakt werden, waartoe Ik geene vrijheijd hebbe. De voornaamste Baaijen sal ik in groot bestek er bij voegen, tot narigt voor de Zeevaart, die onse Cust gansch onbekend is." (Schutte 1982, pp. 109–110). With the words "waartoe ik geene vrijheijd hebbe" Van Plettenberg refers to his doubts about the right to take military action against the Xhosas if they crossed the border. He seems to argue that he would have that right after the Gentlemen XVII had approved the map.

[27] During the term of office of governor Plettenberg parts of the Great Fish River were proclaimed the eastern border after negotiations in 1778. This was reconfirmed by the Council of Policy in 1780. Nevertheless, uncertainty about the borders of the Cape Colony persisted (Bergh & Visagie, 1985). C. Koeman included a map by Leiste (Koeman, 1952b, Pl. VI). Generally speaking, political boundaries in Europe were also still unclear in the eighteenth century and fairly impermanent. Attempts to obtain legal certainty were made by jointly drawing up maps for treaties (Konvitz, 1987, pp. 32–33). In South Africa this was not possible.

Plettenberg's successor, C.J. van de Graaff (governor from 1785–1791).[28] Only in 1798, during the first English occupation, did the Colony get a closed border, which was then indicated on maps.[29] Governor Macartney proclaimed that the border of the Cape Colony ran from the mouth of the Fish River via a boundary beacon erected by Van Plettenberg in the north-east in 1778 up to the mouth of the Buffels River in the north-west.[30]

Unlike the route maps produced by expeditions and Gordon's large map of southern Africa, which went as far as Gordon's knowledge allowed, the official topographic maps produced after 1778 were mainly of administrative significance.[31] They show almost exclusively the territory of the colony and in that way they constitute it. While colonial officer bearers collected information about the country and the inhabitants of the Cape Colony in an unsystematic manner during inspection trips and, judging by Van Plettenberg's doubts, vague agreements were made with the Xhosas, the territory of the colony was by and large coloured in on the topographic maps. This suggested a total overview and legal certainty. Maps were in a general sense a convenient means of legitimising claims to colonial territory.[32] By the end of the eighteenth century the topographic maps gave the colony a visual shape, which could form the self-evident basis for administrative practice.[33] The most important and most dangerous aspect of these maps was that they created an illusion of certainty about the border that separated the colony from 'Kafferland' in the east. The border of the colony acquired an

---

[28] Koeman (1952b, p. 25). The map is included in Koeman's atlas as Pl. XI (Koeman, 1952b).

[29] For example on the map included in Barrow's *Travels* (1801) and a map drawn in 1805 during the Batavian period by J.H. Voorman ("Generale Kaart van de Colonie Kaap de Goede Hoop in Zuid-Afrika [...]" (General map of the Colony Cape of Good Hope in South Africa), CA, M 4/9).

[30] See Macartney's proclamation of 14 July 1798 (Theal, 1905, pp. 450–452). See Penn (1993) for the maps drawn by Barrow during the first English occupation.

[31] Compare the large map by Gordon (Rijksprentenkabinet, Gordon Atlas, G 3).

[32] Harley (2001, p. 165); MacMillan (2003, pp. 426–430).

[33] According to Zandvliet (1998, p. 132) the VOC attached great value to maps as the basis for an efficiently organised administrative apparatus. Jacques Revel (1991, p. 147) makes the following observation in an article in which he discusses among other things the political importance of the cartographic shape of France: "[...] what better symbolizes the rulers' hold over the territory, what better expresses control over it than the map. [...] It is from the map on the classroom wall that children have learned, over the decades, to read the contours of their country, have learned of the unbearable loss of Alsace and Lorraine after the defeat of 1870 [...] Reduced to a hexagon, the national shape became both evident and necessary".

ontological meaning for its European administrators, but meant little to the semi-nomadic Xhosas, who did not know topographic maps and had no central government.

The administrative exploration network confined itself to colonial territory.[34] By acquiring knowledge, it was hoped that the territory could be governed better. Cartographic knowledge ensured that the colony became a unit and could be divided into clearly demarcated districts.[35] Maps helped the colonial administrators to get a symbolic grip on their territory.

Scientific exploration networks

The customary way of collecting knowledge outside the colonial territory was to embark on a reconnaissance journey, which yielded a chronological travel account and a route map as end products.[36] In its expeditions, the VOC administration played the part of a meddlesome "centre of calculation" that determined what information explorers had to collect and how they had to act. By binding the conduct and observations of travellers to rules as much as possible, the travellers would ideally act as a telescope in the hands of their principals. In France this system was increasingly also extended to scientific travellers.[37] In South Africa scientific travellers acted independently, even if, like Alberti, they sometimes made use of guidelines drawn up by armchair scholars to sort the information they had collected.[38]

As stated earlier, the colonial administration could make good use of employees with scientific capacities. The expertise of Barrow and

[34] Between 1721 and 1730 this sort of mapping was also done at Delagoa Bay. As an exception, J.H. Voorman's "Generale Kaart van de Colonie Kaap de Goede Hoop" of 1805 (CA, M4/9) also contains information about the area outside the borders of the Colony, obtained by an expedition to the Tswanas in 1801–1802. Most of the topographic maps from the period under discussion were limited to the territory of the colony.

[35] This can for example already be seen on Leiste's maps, where in the east the boundary between the districts of Stellenbosch and Swellendam are clearly shown.

[36] Edney refers to 'reconnaissance' here, which he defines as "the purposeful movement through and examination of the world" (Edney, 1999, p. 176).

[37] Kury (1998); Spary (2000, p. 84).

[38] One exception was Francis Masson, who received his instructions from Joseph Banks in London (Drayton, 2000, p. 46; Fulford, Lee & Kitson, 2004, pp. 90–92). Within the discourse on South Africa, Masson played a part mainly in the field of botany.

Lichtenstein was used for mapping out the Cape Colony. It was probably partly owing to the anthropological knowledge acquired by Alberti that the eastern border of the Cape Colony remained at peace in the Batavian period. The above-mentioned travellers in the service of the colonial government also had their own ambitions, however. Because of their employment relationship they formed part of the administrative network of the Cape Colony, but their ambitions connected them to the international scientific world.

Lichtenstein had initially come to South Africa as part of governor Janssens' entourage as a private tutor because he had wanted to visit South Africa since his childhood.[39] Barrow also seems to have had his own reasons to take up Governor Macartney's request to undertake a journey of inspection through the colony. Robert Gordon was most skilful at using his official position for his scientific interests. To date no indications have been found that he had been officially instructed to make his long journeys, although he must have had permission to leave the Cape garrison for such a long period. His research as naturalist and ethnographer can have only been of limited use to the VOC administration, as the VOC was mainly interested in the economic potential offered by the interior. At best, Gordon's travels gave the administration a better view of the geography of South Africa.[40] The limited usefulness of his expeditions to the VOC may also have resulted in Gordon having to defend himself against allegations that he had only come to South Africa to do scientific research.[41]

Peter Kolb is a somewhat different case. He had come to South Africa by order of a Prussian *Geheimrat* to make astronomical observations, but he hardly mentions this in *Capvt Bonae Spei Hodiernvm.* While he was secretary of the VOC in Stellenbosch he had access to information from the interior at the local magistrate's office and made friends with the former secretary of the Council of Policy, Van Grevenbroek (1644–1726), who had collected ethnographic information about the Khoikhoi.[42] The interests of scientific travellers combined very well with the interests of the administrative network.

---

[39] Lichtenstein (1811, p. 5).

[40] For this the VOC was not dependent on Gordon, because the surveyor Leiste was drawing his maps of the Cape Colony at the same time.

[41] National Archives (NA), Fagel-Archief, nr. 2515.

[42] In 1695 Van Grevenbroek had written a letter, not published in his time, in which he paints a positive picture of the Khoikhoi (Van Grevenbroek, 1933; Van Stekelenburg, 2001); see chapter 2.

Private travellers such as Le Vaillant, Sparrman and Thunberg also made use of the colonial system. Le Vaillant, for example, already received support from VOC director Jacob Temminck in the Netherlands and later from vice-governor Hacker at the Cape and from the fiscal Boers, recorded as corrupt, who took a personal interest in the natural history research.[43] Among other things, Boers financed new equipment for Le Vaillant after he had lost his own in a battle between Dutch and English ships during the fourth Anglo-Dutch War in Saldanha Bay. Le Vaillant used Boers' house in Cape Town, Rust en Vreugd, as a depot for his collection after his first journey (fig. 44). In order to reward Boers for his assistance, Le Vaillant dedicated his *Voyage* to him in 1790.

The above warrants the conclusion that various forms of symbiotic relationships existed between scientific research and the colonial establishment in the eighteenth-century Cape Colony, from which the scientists, however, were the main beneficiaries. This can also be seen from the publication history of some travel accounts from this period. Lichtenstein's *Reisen* relied on information he had collected as a colonial employee. Large parts of it had even been taken from the official travel journals of Janssens and De Mist. The VOC travel journal of Carel Frederik Brink of 1761–1762 was published as a scientific text by Professor Allamand. Travel journals that had been written with commercial or administrative aims contained factual material that was regarded as interesting from a scientific point of view, even if at the end of the eighteenth century it sometimes became necessary to increase the scientific weight of texts of this kind, as appears from Allamand's reworking of Brink's travel journal.

In principle, scientific travellers in South Africa acted independently from one another. It was purely incidental that, for example, the two students of Linnaeus, Sparrman and Thunberg, travelled through the interior together and that Gordon was accompanied by the Englishman William Paterson, but this took place in a collegial relationship, with each writing his own travel journal.[44] It is for this reason that I want to refer to scientific exploration networks in the plural. The relationships between the travellers were more at the level of a shared scientific practice: they read the works of their predecessors, attempted to

[43] Le Vaillant (1790, pp. 1–2, 28).

[44] Collegial interaction does not mean that the authors shared the same opinions. Berman (1998, pp. 21–64), for example, shows how the descriptions by Georg Forster and James Cook, which deal with the same events, differ from one another despite each gentleman holding the other in high esteem.

obtain certainty about the same issues as their predecessors (and added new ones as well), used the same empirical research method and saw themselves as fieldworkers in contrast to the armchair scholars who had never left Europe. It is because of this shared scientific practice that the scientific South African travellers create, in the terminology of the Polish sociologist of science Ludwik Fleck, the impression of a "Denkkollektiv",[45] like scientists in modern disciplines.

It was a central tenet of the epistemology of eighteenth-century scientific travellers that the facts described also had to be observed with one's own eyes. Reliable knowledge should be based on *autopsy*. The visual observations were captured in notes of all kinds, but also by making drawings of plants, animals, people and landscapes on the spot. Moreover, the observations had to be meticulous, which especially in Gordon's case resulted in the collection of all sorts of data such as dimensions of animals, barometric pressure and temperatures, which were obtained by means of instruments.[46] Typical of the trouble that Gordon took to make accurate observations is the following statement: "not to forget a small trickle of water running along the indentation in the rock, almost against its nature."[47] Even the smallest detail deserved concentrated attention.

Incidentally, within geography the emphasis on own observation was as old as the field of knowledge itself. Herodot regarded autopsy as the most important means of acquiring knowledge.[48] Early modern empiricism subsequently elevated sensory knowledge, especially visual observation, to an epistemological principle of scientific practice and used it as an antidote to belief in traditional ideas.[49] In order to be credible, however, the observer had to be reliable as well, which was guaranteed by his social status as a gentleman or by his special exper-

[45] I use Fleck's original German term, which in the English version is translated as 'thought collective' (Fleck, 1979, pp. 98–110).

[46] During the eighteenth century accuracy in the production of representations (maps, descriptions, pictures) was increasingly emphasised, which resulted among other things in measuring instruments playing an increasingly important part (cf. Licoppe, 1996; Bourguet, Licoppe & Sibum, 2002).

[47] "het aflopen van een kleine streep water langs de verdieping der rots, byna tegen syne natuur niet te vergeten" (Brenthurst Library, Houghton, Gordon Papers, MS 107/15, Miscellaneous notes).

[48] Hartog (1980, p. 272).

[49] Shapin (1996, pp. 69–70).

tise.[50] In any case, he should not lie. Anyone who, like Le Vaillant, did not take truth too seriously and had not experienced the things he recounted therefore attracted the anger and ridicule of other members of the *Denkkollektiv*, such as Gordon and Barrow.

From the end of the eighteenth century, autopsy gained a cartographic foundation in the published travel accounts about South Africa as well. These books contained a map on which the route followed was indicated by a dotted or continuous line, usually with an indication of the names of the places where the traveller had set up camp (fig. 46). By indicating the names of the camping sites it was possible to correlate the chronological travel account with the map.[51] The effect of this correlation is that the observations in the text have a foundation in the trail on the map, which in turn constitutes a reflection of the space of which it is an image. The addition of a map to a travel account made the observations "with own eyes" even more convincing.[52] Each observation could now be located approximately on the map, which constitutes an image of the area represented as closely as possible. The cumulative reality effect of these links is that the observations appear to be *almost* directly in the space represented.[53] Almost, because somebody like Lichtenstein does not conceal the fact that his map is a complex construction that only approximates reality and lays no claim to completeness.[54]

---

[50] I oversimplify slightly here. In fact a controversy exists in the historic literature about the requirements set for scientific evidence. The historians of science Steven Shapin and Simon Schaffer strongly emphasised status (being a gentleman) as a requirement for witnesses to be held as reliable during the scientific revolution in the seventeenth century (Shapin & Schaffer, 1985; Shapin, 1994). On the other hand Barbara Shapiro (2000, p. 141) says that expertise was held as equally important.

[51] In this regard Lichtenstein says: "Da unsre Karte hauptsächlich dazu dienen soll, die Beschreibung des auf der Reise genommenen Weges zu versinnlichen, so sind alle in dem Buche genannten Örter sorgfältig auf die Reiseroute eingetragen und selbst die mit aufgenommen worden, die, wenn man nicht diese Absicht vor Augen hatte, zu unbedeutend gewesen sein würden" (As our map is mainly intended to show the route of our journey, all places mentioned in the book have been carefully entered on the travel route and even those are included which, if one did not have this intention, would have been too insignificant (Lichtenstein, 1811, I, p. 681).

[52] Lestringant (2003, pp. 12, 266, 353) points out the rhetorical effect of a claim of autopsy. Writers of fictitious travel tales therefore used it to render their stories more plausible.

[53] The term 'reality effect' is derived from Roland Barthes' essay on the "effet de réel" (Barthes, 1968). The reality effect creates the illusion that the representation is reality itself. According to Barthes, this is done by not having the *signifiant* refer to a *signifié*, but directly to reality.

[54] Lichtenstein (1811, I, pp. 673–682).

Apart from epistemology, the travellers also shared the issues they wanted to investigate. On the one hand these were familiar subjects from older travel texts which they wanted to check once more. For example, among Gordon's manuscripts there is a "sheet with questions as regards Kolb", a list of issues arising from Kolb's *Capvt Bonae Spei Hodiernvm*, probably the Dutch translation of the book. Gordon also added answers to many questions, such as the well-known question about testicle removal: "The testicle cutting, was this ever a custom? Answer: Not among the Little Namaquas." This is followed by more specific questions—"How is this done? Is the cutter of the kraal highly respected and what is he called?"—questions to which Kolb gives very precise answers, but which Gordon clearly does not trust.[55] Besides long-standing questions, such as monorchy among the Khoikhoi, there were also new ones. The topic of the day concerned details of the giraffe after the (re)discovery of this animal in 1760. In order to see these impressive animals and to obtain their skeletons and skins, Gordon (certainly) and Le Vaillant (probably) crossed the Orange River into the present-day Namibia. For a long time giraffes had only been known from ancient treatises on natural history. A giraffe was presented to Lorenzo de' Medici by Al-Ashraf Kait-Bey, the sultan of Egypt, in 1486, and a second giraffe to Charles x, king of France, in 1827. The 1486 giraffe was eulogized by poets, but not described scientifically.[56] Hendrik Hop's expedition had collected the first reliable particulars about the animal in 1761–1762. Allamand had included a drawing made by the surveyor Brink in the Dutch edition of Buffon's *Histoire Naturelle* in 1776 and republished it in his edition of the *Nieuwste en beknopte beschrijving* in 1778. Gordon carried out the most complete investigation and also saw to it that a skeleton of the giraffe was shipped to the cabinet of natural history of stadholder Willem V. After the French invasion of 1795 this skeleton and other valuable items from the stadholder's collection were transferred to Paris.[57] In 2007 Gordon's giraffe was still

[55] Answers to the questions mentioned are absent from Gordon's questionnaire, because the answer to the main question—is a testicle cut out—was already negative (Brenthurst Library, Houghton, Gordon Papers, ms 107/6, "Particularités relatives a quelques hordes Hottentottes" (particulars of some Hottentot groups). In the folder Brenthurst Library, Houghton, Gordon Papers, ms 107/12, "Observations & Reflexions sur divers sujets" (Observations and reflections on various subjects) there is also a sheet with notes taken from voc travel journals.

[56] Lagueux (2003), Ringmar (2006).

[57] Rookmaaker (1983; 1989, pp. 291–293).

part of the permanent exhibition of the Musée d'Histoire Naturelle (just behind the visitors' entrance).

Giraffes were spectacular animals to eighteenth-century Europeans. Georg Forster, on a visit to the collection in The Hague, remarked that giraffes were so unusual that they seemed rather to have originated from the imagination than to be a link in the chain of life forms.[58] The enthusiasm aroused by the discovery of giraffes in South Africa gave rise to the suspicion that there might be further sensational discoveries to be made. It led, for example, to renewed searches for another animal that was known only from the literature and that enjoyed a very dubious scientific reputation at the end of the eighteenth century: the unicorn.

I want to dwell on the searches for unicorns a little longer, for they reveal a few important characteristics of the practice of science in eighteenth-century South Africa. The searches made three things clear. Looking for unicorns diverted an important part of the Cape *Denkkollektiv* from the scientific mainstream in Europe, where in the course of the eighteenth century interest in the question of the unicorn's existence was waning. Furthermore, this case illustrates how scientific travellers succeeded in involving persons from outside the world of science in their research project, so much so that the colonial authorities at some stages saw the capture of unicorns as their own task. Lastly, this case also says something about the mentality of the scientific traveller in South Africa, because the desire for new knowledge sometimes made him willing to forget prevailing principles for proper scientific research. The principle of autopsy was sometimes subordinated to the desire for successful publicity. Below I will first describe the origin and evolution of unicorn research at the Cape and then return to the issues mentioned here in the conclusion of the chapter.

---

[58] Forster says the following about his visit in his travel account *Ansichten vom Niederrhein* of 1790: "Herr Vosmaer führte uns freundschaftlich zu verschiedenenmalen in diesem reichen Tempel der Naturwissenschaft umher, und zeigte uns auch die neu hinzu gekommene Stücke, die noch nicht an ihrem bestimmten Orte aufgestellt waren, wie [...] das Gerippe des Cameleopardalis der Alten oder der Giraffe der Neuern, dieses seltsamen Thieres, das mehr einem Traum der Einbildungskraft, als ein Glied in der Naturkette ähnlich sieht" (Mr Vosmaer kindly took us on a tour of this rich temple of natural history [the stadholder's collection in the Hague] several times and also showed us the newly arrived items which had not yet been set up in their place, such as [...] the skeleton of the Cameleopardalis of the ancients or the Giraffe of today, this remarkable animal that looks more like a dream of the imagination than a link in nature's chain (Forster, 1967–1970, II, p. 715)).

Unicorns in South Africa

In 1797 John Barrow found a picture of a unicorn. It had not been easy. After a difficult trek from cave to cave in the Tarka Mountains in the east of the Cape Colony, his entourage had reached a high and densely vegetated gorge. Right at the back of the gorge was a cave, its opening closed off by shrubs. One of Barrow's companions wrestled through the bush and shouted to the others that the walls of the cave were covered in Bushman drawings. After the shrubs had been removed, Barrow was able to inspect the drawings. He had finally found a picture of a unicorn, although unfortunately it was no more than a head, because an elephant had been drawn over the rest of the body. Barrow copied the drawing and published it later in his travel account (fig. 45). He also noted the location of the cave on his map of the Cape Colony (fig. 46). The head of the unicorn was the first tangible evidence of the existence of the South African unicorn. Nevertheless, Barrow was disappointed that he had not found a complete picture. His disappointment aroused the amusement of the Cape colonists who had accompanied him on his search. It clearly did not matter to them whether the unicorn was found. However, the moment Barrow promised a reward of 5000 rix-dollars for anyone who "would bring him an original" the amusement turned into enthusiasm. They were immediately ready to join an expedition to the country behind the Bamboo Mountain, beyond the borders of the colony, the area in which unicorns would be found.[59]

The search for unicorns in South Africa continued from 1776 until deep into the nineteenth century. What is remarkable about this is that in southern Africa searches were launched, while the scientific mainstream in Europe had long stopped believing in the existence of unicorns. Buffon paid no attention to the unicorn in his very complete *Histoire Naturelle*, nor did Linnaeus in his *Systema Naturae*.[60] In the past only two studies were dedicated to the South African searches, but they

[59] Barrow (1801, pp. 311–313).

[60] Shepard (1967, p. 204). Lorraine Daston and Kathrine Park (2001, pp. 331, 333) mention this as an example of the dwindling faith in wonders in the eighteenth century, which they view as a key development in the science of the Enlightenment. At present the dissertation by Faidutti (1996) is the most complete overview of the European debate about the existence of these animals up to the nineteenth century.

have left no imprint in the more recent literature.[61] A pioneering work was an article by Smith in 1968 in which an inventory of the passages in travel accounts dedicated to unicorns was drawn up. An article by Voss in 1979 relies heavily on Smith's article. The main addition Voss made to the philological unicorn research in South Africa was a classification of the explanations given for the existence of the unicorn:

1. The unicorn is a man-made animal (an animal of which one horn has been removed);
2. the unicorn is an animal that did exist but has become extinct, or an animal that lives in *terra incognita*;
3. the unicorn is actually a rhinoceros or an oryx;
4. lastly, of course, that it is a mythical animal.

It was Anders Sparrman who turned the unicorn into a subject of scientific research in South Africa. In 1783 Sparrman was the first to devote a number of pages of his travel account to the unicorn. Like Barrow, he had heard about rock drawings of unicorns in the eastern part of the colony. His informant was a colonist named Jacob Kok, from Agter Bruintjes Hoogte. That there were such drawings to be found somewhere was sufficient reason for Sparrman to believe in the possibility of the animal's existence. After all, it was impossible for the "Snese Hottentotten" (literally: Chinese Hottentots, or San) to have thought up the unicorn; for that they were too primitive. As a primitive people they consequently had to reflect reality in their rock paintings, so that a rock painting of a unicorn proved that such an animal had indeed been seen by the Bushmen. Then there was a second argument: South Africa and its fauna were to a large extent virgin territory to eighteenth-century naturalists. There was so much that was still unknown. It had only recently been discovered how elephants mate, and when Sparrman wrote his book the giraffe had only just been promoted from an almost mythical beast to a zoological species. Why could the same not happen to the unicorn as to the *camelopardalis*, until recently

[61] Smith (1968) and Voss (1979). The searches for unicorns in South Africa have received little attention in the international literature. Faidutti (1996), who wants to give a complete overview of the discourse on unicorns, does not refer to these articles, but does briefly cover the South African searches in chapter 3.3 of his dissertation. Gerritsen, however, says nothing about the matter of the South African unicorn in his exhibition catalogue when he suffices with the remark: "[i]n the eighteenth century the scholarly debate about the unicorn falls silent" (Gerritsen, 2003, p. 47).

deemed equally fantastic?[62] To support the statement that the unicorn did exist, Sparrman printed a fragment from a letter by the German explorer of Siberia and eminent member of the Russian Academy of Sciences and the Royal Society, Peter Simon Pallas (1741–1811), in which Pallas regards the existence of unicorns as possible. Support from an authoritative zoologist bolstered Sparrman's claim.

After Sparrman had devoted a few pages of his travel account to the unicorn, the animal continued to tickle the curiosity of subsequent authors. The problem of the South African unicorn even became a topos in documents on the interior of South Africa. The fact that this was possible is related to the nature of scientific fieldwork in the eighteenth century. In their books travellers were constantly busy comparing their own observations with the descriptions given by their predecessors. Once a certain subject had been introduced by an authoritative author, such as the learned Sparrman, a pupil of the great Linnaeus, with the apparent support of Pallas, it stubbornly kept turning up in the writings of others. Since the influential Barrow also devoted much attention to the animal, the unicorn had become firmly established in the zoological discourse on South Africa.

It is curious that the searches for unicorns only started in 1776, with Sparrman's journey. At the beginning of the eighteenth century Valentyn mentions the unicorn,[63] and there are a few more authors, but nobody found it necessary to fill his pages with theories on the possibility of the existence of a unicorn in South Africa until Sparrman did it. It is remarkable that the unicorn only put in an appearance in travel documents on South Africa at the end of the eighteenth century, after it had not drawn any more attention than cursory remarks before. One would have expected that the interest would have slowly decreased when the empirical study of natural history began to dominate scientific research. In the course of the Enlightenment more and more miracles were banished from the field of science. The horns of unicorns in curiosity cabinets had already been reclassified as narwhal horns long ago.[64] What is noteworthy about the situation in South Africa is that the unicorn owed its new lease on life precisely to modern science.

---

[62] Sparrman (1977, II, pp. 116–118).

[63] Valentyn (1973, p. 118).

[64] Daston & Park (2001, p. 332). Daston and Park may have been a little too definite in their allegations about the eighteenth-century cleaning-up action with regard to miracles, as will appear further below.

It is well known that the unicorn has a long history. The Greek author Ktesias (around 400 B.C.) introduced it into the scientific literature, locating it in India. Aristotle (384–322 B.C.), and Pliny (23–79 A.D.) quoted him on this. Julius Caesar (100–44 B.C.), on the other hand, believed that the unicorn was found in Germany. He said that in this impenetrable country there were dense forests with many as yet unknown animals, including the unicorn. The unicorn also played a part in religious texts. The Hellenist translators of the Septuagint had seen to this by translating the Hebrew word *re-em* with the Greek word for unicorn, *monoceros*. The Church Fathers subsequently polished up the unicorn allegorically as a symbol of Christ. The scientific and the theological tradition were then fused by the *Physiologus* (about 300 A.D.) into allegorical tales of animals where the unicorn (Christ) is captured by a virgin (Mary). This subject is richly represented in mediaeval iconography.[65]

In the eighteenth century this tradition encountered a healthy and well-founded distrust. According to the philological unicorn expert Shepard, the unicorn was in a bad state before Sparrman brought him out of the stable.[66] Whoever published anything on South Africa took little interest in the unicorn up till the end of the eighteenth century. He is not present in Peter Kolb's encyclopaedic work *Capvt Bonae Spei Hodiernvm*, which attempts to be as complete as possible. South Africa first had to become the hunting ground of the eighteenth-century science of nature, natural history. An important aspect of scientific observation was comparison with the existing literature. Sparrman, for example, wanted to know more about the mating behaviour of elephants. He knew from the literature that little was known about this because the elephant was too chaste to perform carnal acts in public view, even in a dark cage. For this reason alone the story that the female elephant lay on her back while mating was suspect. In South Africa, Sparrman heard that the female went down on her front knees to give the bull access. His informer had seen a female wait like this for two hours while two bulls fought for the honour. Tired of waiting, the informer had left before the act was consummated. But this was enough for Sparrman to contest the current opinion.[67] After all, the instruction

[65] Shepard (1967); Freeman (1976); Beer (1977).
[66] Shepard (1967, p. 204); Thomas (1984, pp. 79–80).
[67] Sparrman (1977, II, p. 297).

of natural history was, according to Sparrman's tutor Linnaeus, to unmask superstition.[68]

Eighteenth-century science was not always consistent, however—a fact that is sometimes overlooked in the recent historiography of science during the Enlightenment.[69] When in 1783 the German philosopher Georg Christoph Lichtenberg (1742–1799) gives a summary of the achievements of which the eighteenth century can rightfully be proud, he also mentions the discovery of animals that are even more wondrous than the fabled 'Lema snake'.[70] Even a person with the status of Linneaus could look for confirmation of traditional knowledge or even uncritically believe it. When a 'mermaid' was caught in Denmark, Linnaeus was greatly interested. In a letter to the Swedish Academy of Sciences he solemnly wrote that science was still unsure of the existence of such a creature. On another occasion he proved to be a supporter of the view that swallows hibernate on the bottom of a lake, although this is anatomically impossible.[71] And if one has a look at Linnaeus' primates, one will find besides *Homo sapiens* also *Homo caudatus* ('tail man') and the troglodyte there.

Against this background it was not strange that his pupil Sparrman hoped to find a unicorn in South Africa. Within the "style of reasoning"[72] of eighteenth-century natural history, the reservoir of knowledge consisted, besides empirically confirmed data, also of book knowledge that could be confirmed or refuted, but could not be consigned to the realm of fables without further ado, despite the emphasis on autopsy. Moreover, many curious beasts were wandering about in Africa of which little or nothing was known. Sparrman mentions one of them when he defends the possible existence of the unicorn: the giraffe. Like the unicorn, the giraffe had for a long time been known only from the literature. The fact that Sparrman dwelled on the unicorn for so long in his travel account has in my view much to do with the recent discovery of the giraffe, which had until only shortly before—also according to Sparrman—been considered a mythical animal. If such a curious animal as the giraffe was found to be living in southern Africa, why

[68] Broberg (1983, p. 177).

[69] The latest literature strongly emphasises that the science of the Enlightenment wanted to put a stop to faith in wonders (cf. Daston & Park, 2001, pp. 329–364).

[70] Quoted in Bierbrodt (2000, p. 24).

[71] Broberg (1983, pp. 177–178).

[72] Hacking (1982, pp. 48–66).

not the unicorn? After all, we know so little of South African nature, Sparrman argues. Have we not only just found out how the well-known elephant mates?[73] For as long as hard proof was absent, the existence of the unicorn remained a debatable subject.

However, the most empiricist of South African researchers, Robert Jacob Gordon, rejected the existence of the unicorn.[74] In Europe Sparrman's remarks were criticised by the renowned professor Petrus Camper (1722–1789) of Groningen. In 1786 Camper commented devastatingly in the *Schriften der Berlinischen Gesellschaft* on Sparrman's assertions about the unicorn. What Sparrman had to say about the unicorn "did not amount to much", and to cap it all his information came from witnesses who "as far as science is concerned, are in the greatest state of ignorance". Camper then proves by way of comparison with other, known horned animals that the unicorn is an anatomical impossibility. The main element in his argument is that all horned animals have two horns which can only be located on both sides of the head. The horn of the unicorn would therefore grow in an impossible place, on the joint where the two parts of the bone of the forehead (*os frontis*) come together. Close by, at the point where the unicorn's horn would have to be, there are also cavities indispensable for breathing (the *sinus frontalis*) that allow the passage of air. Anatomically seen, there is consequently no room for a horn in the middle of the skull of a mammal. His conclusion is that the unicorn is a mythical animal that owes its existence to a painter or an "idolater" who had thought up an allegorical animal in an Egyptian way. As far as he is concerned, the unicorn therefore belongs to the ranks of other proven mythical creatures such as the seven-headed dragon of Seba, sirens and Rondelet's bishop and monk covered in scales.[75]

Against the background of general—and, in the case of Camper, well-founded—scepticism those who defended the existence of the unicorn had to underpin their suppositions with arguments, in which nobody actually took account of Camper's anatomical objections or

---

[73] Sparrman (1977, II, p. 117).

[74] Sparrman (1977, II, p. 116, nt. 35).

[75] Camper (1786, pp. 219–226). Camper is referring to the seven-headed snake in the *Thesaurus* (1734–1735) of the Amsterdam apothecary Albertus Seba (1665–1736) and the so-called sea bishop from the *Libri de Piscibus Marinis* (1554–1555) by Guillaume Rondelet (1507–1566). Faidutti says that he was unable to find Camper's article (Faidutti, 1996, chapter 3.3). Barrow's report on the unicorn in his *Travels* also attracted criticism in a German journal (Von Knebel, 1802, pp. 94–100).

paid much attention to the status of the witnesses who said that they had seen unicorns. These are usually the following arguments:

1. Rock paintings by San show unicorns. Because these people are too primitive to imagine mythical creatures, the depictions must correspond to reality.[76] Nevertheless, the only one who succeeded in reproducing a drawing was Barrow.[77]
2. Informers—colonists or aborigines—have seen the unicorn. According to Lady Anne Barnard, the wife of an English colonial official at the Cape, the "Hottentots" recognized the unicorn when they saw it in the British coat of arms.[78] The Cape freeburgher Hendrik Cloete had heard the tale about unicorns from the "Bastaard Hottentot" Gerrit Slinger. During an expedition beyond the eastern border Slinger had encountered a herd of unicorns of which he was able to give a description.[79] Barrow included descriptions of unicorns by three colonists in his travel account.[80]
3. The existence of a unicorn could also be made plausible by evidence from the literature and authoritative authors. Barrow refers to the description of unicorns by the seventeenth-century Jesuit Lobo[81] and Sparrman adds a passage from a letter from the authoritative German naturalist Pallas that confirms the possible existence of the unicorn on the basis of old travel accounts.[82]

Barrow formulated the scientific problem around the question of the existence of the South African unicorn concisely from the perspective of the travellers:

> The schooled mind is apt to feel a propensity for rejecting every thing new, unless conveyed to it through the channel of demonstrative evidence, which, on all occasions, is not to be obtained; whilst, on the other hand, credulity swallows deception in every flimsy covering. The one is, per-

---

[76] Sparrman (1977, II, p. 117).
[77] Barrow (1801, p. 313).
[78] Anderson & Lloyd (sa, p. 22).
[79] Schutte (1982, pp. 296–297). Cloete's letter was published by K.K. Reitz in the *Verhandelingen uitgegeven door het Zeeusche Genootschap der Wetenschappen* in Flushing in 1792 and later published in translation in the *Magazin für das Neueste aus der Physik und Naturgeschichte*, Band 10, 3. Stück, 1796, pp. 64–72). I was not able to inspect the Dutch text of the article by Reitz.
[80] Barrow (1801, pp. 318–319).
[81] Barrow (1801, p. 316); Lobo (1728, pp. 69–70, 230–231; 1984, p. 166).
[82] Sparrman (1977, II, pp. 117–118).

> haps, equally liable to shut out truth as the other is to imbibe falsehood. Nature's wide domain is too varied to be shackled with a syllogism. What nations, what animals, what plants, and other natural productions, may yet be discovered in the unknown parts of the globe, a man, who has studied nature in the closet only, would hardly be supposed presumptuous enough for conjecture [...].[83]

As long as the world had not been investigated in its entirety, the unicorn could exist—somewhere beyond the eastern borders of the Cape Colony, "between the so-called table mountain and the zeekoe river, behind the Bamboes Mountain", somewhere "behind Tamboekies and Cafferland" (according to the Cape colonist Hendrik Cloete).[84] For the same reason people began looking for unicorns in Tibet in the nineteenth century.[85]

The discourse on the South African unicorn was in any case a product of modern natural science and cannot be traced back to popular culture in the Cape Colony. One could say that the unicorn was introduced into South Africa from Europe. The greatest contribution in this regard was made by Sparrman and Barrow. Within the discourse on South Africa they suggested the possibility that the South African unicorn could exist. The inhabitants of the Cape Colony were only stimulated to take part in the research after money prizes had been offered. We already saw this happening when Barrow offered a prize of 5000 rix-dollars. During the Batavian period governor Janssens and commissioner-general De Mist raised the amount to the price of an ox wagon with a team of 12 oxen. Janssens was even prepared to add another 1000 rix-dollars.[86] The colonial government therefore also saw the scientific search as its own project.

Conclusion: scientific practice at the Cape

The history of the searches for the South African unicorn reveals a scientific attitude that partly deviates from what was generally acceptable in Europe at the same time. Two aspects in particular are striking. First,

---

[83] Barrow (1801, p. 317).
[84] Schutte (1982, p. 420); Barrow (1801, p. 315); Paravicini di Capelli (1965, pp. 145–146).
[85] Faidutti (1996, chapter 3.3).
[86] Paravicini di Capelli (1965, pp. 145–146); Molsbergen (1916, pp. 174, 224); Lichtenstein (1811, I, p. 269).

that the belief in the existence of a unicorn in South Africa was fanned and kept alive by indirect evidence. When the researcher was unable to make an observation himself, it sufficed that all sorts of colonial witnesses who were less reliable within the current epistemology—Boers, Bastaards and San—had observed the phenomenon.[87] Second, the scientific ratio could be turned against itself in the absence of hard evidence, as the quotation from Barrow shows. Absence of proof was no proof of absence. To be precise, the denial up front was irrational. Barrow's argument reminds one of the anecdote told by John Locke (1632–1704) about the discussion between the King of Siam and the Dutch ambassador. The Dutch ambassador told the King, among other things, that the water in his home country became so hard in cold weather that you could walk on it. To this the King replied: "Hitherto I have believed the strange things you have told me, because I look upon you as a sober fair man, but now I am sure you lie."[88] When it came to unicorns, one should not make the same incorrect inference as the King of Siam.

The case of the unicorn reveals a few more peculiarities about what we will simply call Cape science. First of all, there was the remarkable success in obtaining support from people outside the scientific network. Scientists managed to find a broad range of colonial assistance for their project, even if it occasionally required a financial incentive. In the Batavian period the government was even prepared to supply the encouragement, and that at a time when the same government had to cope with large financial deficits. The successful execution of scientific research was therefore co-dependent on cooperation with colonists and colonial officials.

It is also noteworthy that the Cape fieldworkers came into conflict with European armchair scholars. In the quotation given above, Barrow makes a derogatory comment about armchair scholars, while Petrus Camper said rather unkind things about Anders Sparrman.[89] In the practice of science during the Enlightenment such conflicts between fieldworkers and armchair scholars were not uncommon. Travellers

---

[87] Knowledge based on hearsay had no validity within the empiricist epistemology (Dear, 1985, p. 152). Moreover, one had to be a gentleman or expert in order to be a credible witness.

[88] Locke (1997, pp. 579–580). The first edition of Locke's *An essay concerning human understanding* was published in 1670.

[89] The French anatomist Georges Cuvier also turned against the existence of unicorns in 1827 in a comment on Pliny (Faidutti, 1996, chapter 3.3).

generally based their authority—though not in the case of the unicorn, obviously—on the direct observation of natural phenomena, while the scholar in his anatomical cabinet or his museum of natural history appealed to the bigger picture because he was able to compare the specimens collected around him in his own good time.[90] According to the latter, knowledge was not born in the field, but in the museum, "the centre of calculation", in Latour's terminology. In this sense, Camper's reaction is characteristic because he refutes Sparrman's assertions by means of an anatomical comparison with other, familiar animals.

The most striking feature remains, however, that by clinging to their belief in the existence of unicorns the Cape travellers simply trampled established epistemological principles underfoot. The assertions about unicorns were not based on autopsy—according to the epistemology of the time, the fieldworker's best means of acquiring knowledge—but only on dubious witnesses. This leads to the suspicion that the desire to discover unicorns must have been very strong. Scientific curiosity is often mentioned by travellers as the reason to make the long journey to South Africa. This was more than a topos.[91] Thomas Hobbes (1588–1679) had already classified scientific curiosity as a desire: "*Desire*, to know why, and how, CURIOSITY [...]" Scientific curiosity was, according to Hobbes, a pure desire characterised by a "perseverance of delight in the continuall and indefatigable generation of Knowledge, [which] exceedeth the short vehemence of carnall Pleasure".[92] When it came to unicorns, this desire was less pure, however, and could not always be kept in check by reason. After all, one could become famous by discovering natural wonders, such as unicorns, in the same way or even more than with giraffes. The giraffe was one of the main reasons why Le Vaillant, eager for publicity, had travelled to the Cape.[93] It was not

[90] For the contrast between armchair scholar and fieldworker in the science of the Enlightenment see Outram (1996, pp. 259–262; 1999, pp. 286–287).

[91] 'Curiosity' was a topos in early modern travel accounts, but it could also be the true motive for the journey, even if the risks were high. Van Gelder is prepared to accept curiosity as a motive for the journey if there were no demonstrable, compelling economic reasons for it (Van Gelder, 1993).

[92] Quoted from Daston & Park (2001, p. 307).

[93] "Depuis long-tems c'étoit l'objet le plus ardent de ma curiosité que ces animaux [i.e. giraffes], si peu connus des naturalistes et même des habitans de la colonie. Un des premiers motifs de mon voyage avoit été de les étudier et de les connoître; et je me seroit reproché toute ma vie d'en avoir l'occasion et de m'y être refuse" (Since long ago these animals [i.e. giraffes], so little know among the naturalists and even the inhabitants of the colony, had been the most intense object of my curiosity. One of

for nothing that Le Vaillant had had himself drawn together with a giraffe on the title page of his *Voyage* in 1790, even when the hunt for this animal only came up in the *Second Voyage*, which appeared in 1795. In the second half of the eighteenth century this curiosity about sensational marvels of nature had become somewhat suspect in scholarly circles. According to the *Encyclopédie*, the "noble" scientific interest was directed at ordinary natural phenomena such as gravity and electricity instead. Scientific curiosity titillated by the wonders of nature, on the other hand, was deemed vulgar.[94]

Granted, the case of the unicorn is an extreme example, but it was precisely this case that revealed certain trends in the Cape exploration business more clearly. Whereas European science in general pursued a demystification of the world view that left less and less room for wonders of nature, the Cape periphery offered its own space for knowledge production where presumed wonders of nature could become scientific reality.[95] An encyclopaedic science that focused on (among other things) amazing phenomena and viewed the foreign world as a colossal cabinet of curiosities could endure here better than in Europe, where science was focusing increasingly on 'normal' natural phenomena.[96]

This interest in the unknown can also be found in the field of ethnography. On the whole, researchers at the end of the eighteenth century were no longer very interested in assimilated Khoikhoi. Their interest in the colonists was mainly determined by politics and aimed at demonstrating the reprehensibility of the Dutch colonial system or of colonialism. Ethnographers were interested mainly in the indigenous peoples living outside the colonial sphere of influence, and they were especially interested in curious habits and unusual physical characteristics. Unlike the unicorn research, this did not lead to clashes with armchair scholars, because ethnographic research had developed as a field science from the outset.[97] The traveller as an ethnographer

---

the main reasons for my voyage was to study and to get to know them, and I would have reproached myself for the rest of my life for having had the opportunity to do so and having turned it down.) (Le Vaillant, 1795, I, pp. 1–2).

[94] Daston & Park (2001, pp. 327–328).

[95] Livingstone (2004, pp. 144–145) refers to the "site of exploration" in connection with the spatial dimension of knowledge that owes its existence to exploration.

[96] This collection-oriented encyclopaedic science is generally regarded as a phenomenon of the seventeenth and early eighteenth century. Cf. for example Bergvelt & Kistemaker (1992).

[97] Cf. Harbsmeier (1995, p. 20).

automatically had authority based on autopsy that he needed to assert only relative to other travellers. Unlike in zoology and botany, which relied heavily on comparative research of specimens in the natural history cabinet, nobody in Europe was able to speak about the culture of foreign peoples with the same authority as travellers who had actually met them. Armchair anthropologists attempted to compensate for this loss to some degree by drawing up instructions such as Degérando's questionnaire for scientific travellers. But Alberti used this questionnaire as a convenient descriptive model, without worrying about Degérando's underlying endeavour to calibrate ethnographic information for further processing by scholars in Europe. The other travellers allowed themselves to be guided rather arbitrarily by the matters their predecessors had deemed noteworthy.

Bearing in mind the prosperous careers of Sparrman (professor), Thunberg (professor), Le Vaillant (member of France's scientific establishment), Lichtenstein (professor) and Barrow (Secretary of the Admiralty) in Europe, southern Africa also offered careers to persons with scientific ambitions besides the possibility of sensational discoveries. They used the colonial network in the Cape Colony as preparation for these successful careers. What is remarkable about the Cape scientific practice is therefore not so much that the colonial government employed scientifically educated individuals, but rather that these individuals knew how to use the colonial network to their advantage. The net result is that science benefited more from this collaboration than the colonial administrative apparatus did.

If we see the scientific travellers as part of a network, the lines of enquiry run in a different direction than Pratt contends. In her analysis Pratt essentially follows the track of the older colonialist historiography of South African explorers by seeing the scientific travellers as trailblazers of the colonial regime. To Pratt they are also in the first place *Pioneer travelers*, to use the title of Forbes' book of 1965. But in Pratt's view they have changed their heroic image for that of a colonial spy with *Imperial eyes*. In my analysis the scientific traveller plays a more independent role. It is true that the results of his research were occasionally used by colonial government, but the traveller set his research objectives himself in a dialogue with the scientific discourse of the day. In the first place he looked for answers to scientific and not to geopolitical or economic questions. Intellectually he therefore remained relatively autonomous with regard to the colonial dispensation, even if he was clever at exploiting it for his own purposes. The relationship between scientific travellers and colonial context was of a symbiotic nature, but in such a way that the travellers benefited much more from this affiliation than the colonial government for administering the Cape Colony. To put it starkly: in the relation between science and colonialism, it was science that was the imperialist party in eighteenth-century South Africa.

ANNEX ONE

INDEPENDENT EDITIONS AND TRANSLATIONS OF PETER KOLB'S *CAPVT BONAE SPEI HODIERNVM* IN THE EIGHTEENTH CENTURY

The main versions of Kolb's book from the eighteenth century are listed below with the title descriptions partly on the basis of *A South African Bibliography to the year 1925* (1979), without any book historical pretences. Remarks concerning differences between the versions are in square brackets. I have consulted the copies in the South African National Library in Cape Town and the J.S. Gericke Libary in Stellenbosch. Versions in collections of travel writing and editions published after 1800 are not included.

[1] *The original edition*

Capvt Bonae Spei Hodiernvm Das is: Vollständige Beschreibung des Africanischen Vorgebürges der Guten Hofnung. worinnen in dreyen Theilen abgehandelt wird/ wie es heut zu Tage/ nach seiner Situation und Eigenschaft aussiehet; ingleichen was ein Natur-Forscher in den dreyen Reichen der Natur daselbst findet und antrifft: Wie nicht weniger/ was die eigenen Einwohner die Hottentotten, vor seltsame Sitten und Gebräuche haben: Und endlich alles/ was die Europaeischen daselbst gestifteten Colonien anbetrift. Mit angefügter genugsamer Nachricht/ wie es auf des Auctoris Hinein- und Heraus-Reise zugegangen; Auch was sich Zeit seiner langen Anwesenheit/ an diesem Vorgebürge merckwürdiges ereignet hat. Nebst noch vielen andern curieusen und bißhero unbekandt-gewesenen Erzehlungen/ mit wahrhafter Feder ausführlich entworffen: auch mit nöthigen Kupfern verzieret/ und einem doppelten Register ersehen, von M. Peter Kolben/ Rectore zu Neustadt an der Aysch. Nürnberg, Peter Conrad Monath, 1719.

[Format: folio. The first German edition is the only edition in which the description is presented in the form of letters.]

[2] *The Dutch translation*

Naaukeurige en uitvoerige beschryving van Kaap de Goede Hoop behelzende Een zeer omstandig Verhaal van den tegenwoordigen toestant van dat vermaarde Gewest, deszelfs Gelegenheit,Haven, Sterkte, Regerings-vorm, Uitgestrektheid, en onlangs ontdekte aanleggende Landen; nevens Een geleerde Beschryving van het Klimaat en Aart van dat Landschap; van deszelfs dieren, visschen, vogelen, planten, kruiden; mitsgaders verscheide wonderen der natuur, daar te Lande ontdekt; Waar by nog komt, een zeer nette en uit eige ondervinding opgemaakte Beschryving van den oorsprong der Hottentotten: vervattende Een merkwaardig Bericht van derzelver Tale, Godsdienst, Levenswyze, zeldzame Overleveringen, Gewoonten, Maniere van Trouwen, Besnydenis, Opvoeding; als mede veele andere kurieuse Waarnemingen aangaande de Zeden van die Natie; den toestant van die Colonie en Europeaansche Inwoners: in gene andere Beschryving van dit Gewest te vinden. Alles, geduurende een lang verblyf aan de gemelde Kaap, na waarheit beschrevendoor Peter Kolb Meester der vrye Konsten, door een voornaam Minister, met Brieven van voorschryving van wylen den Wel-Ed. Gestr. Heere Nikolaes Witsen, Burgemeester der Stad Amsterdam, tot het opmaken van deze, en ook Mathematische en Sterrekonstige Observatien, na de Kaap gezonden, en aldaar geweest Secretaris van Stellenbosch en Drakenstein; tegenwoordig Rector van het Illustre School te Neustad aan den Aysch. Met kurieuse, nieuwe en nodige Kaarten en vele printverbeeldingen opgeheldert en versiert. 2 parts. Amsterdam, Balthazar Lakeman, 1727.

[Format: folio. Virtually complete translation of *Capvt Bonae Spei Hodiernvm* of 1719, with small changes to the text—omission of the letter form—and new, more attractive illustrations. The Dutch translation is the most expensive edition of Kolb's work. There is a list of subscribers in front of the introduction. The book was published in two formats, 34 cm and 42 cm. According to the *South African Bibliography* (p. 7) all editions in the 42 cm format appeared in two volumes, whereas some editions of the 34 cm appeared as two parts in one volume.]

[3] *The English translation in two volumes*

[3.1]

The Present State of the Cape of Good-Hope: or, A Particular Account of the several Nations of the Hottentots: their Religion, Government, Laws, Customs, Ceremonies, and Opinions; Their Art of War, Professions, Language, Genius, &c. Together with A Short Account of the Dutch Settlement at the Cape. Written Originally in High German, by Peter Kolben, A.M. Done into English, from the original, by Mr. Medley. Illustrated with copper plates. London, W. Innys, 1731.

[3.2]

The Present State of the Cape of Good-Hope: vol. II. Containing, the Natural History of the Cape; or, A particular Description of all the Sorts of Animals and Vegetables in that Neighbourhood; as of Beasts, Birds, Insects, Sea- and River-Fish and Flowers. Likewise an Account of the Mineral Productions, and of the Sea-, River-, and Spring-Waters here. Together with some Observations on the Cape-Winds and Air. To which is prefix'd, A Topographical Account of the Colonies there; as, of their Extent, Rivers, Springs, Mountain Roads, Places of Note &c. Written Originally in High German, By Peter Kolben, A.M. Done into English, from the Original By Mr. Medley. Illustrated with Copper Plates. London, W. Innys, 1731.

[Formaat: octavo. The Engish translation is a much abridged version of *Capvt Bonae Spei Hodiernvm*. Abridged versions of the second and third parts of *Capvt Bonae Spei Hodiernvm* are included in volume 1 of *The present state of the Cape of Good Hope* and the first part of *Capvt Bonae Spei Hodiernvm* in volume 2 of *The present state of the Cape of Good Hope*. A second impression of volume 1 was published by W. Innys and R. Manby in 1738. Volume 2 of the English translation has a subtitle that differs from the subtitle in volume 1. The illustrations in both volumes are based on *Capvt Bonae Spei Hodiernvm* and most of them are mirror images. In most of the studies written in English in which reference is made to Kolb, the authors are not aware that the English translation is an abridged version of Kolb's book. This has sometimes led to misunderstanding. Mary Louise Pratt (1992), for example, did not take note of the original text, nor is she aware of the existence of the second volume of the English translation, which causes her to misrepresent the development of eighteenth-century travel accounts.]

[4] *The French translation in three volumes*

Description du Cap de Bonne-esperance; Où l'on trouve tout ce qui concerne l'histoire naturelle du pays; La Religion, les Moeurs & les Usages des Hottentots; et l'etablissement des hollandois. Tiree des memoires de Mr. Pierre Kolbe, Maitre ès Arts, Dressés pendant un séjour de dix Années dans cette Colonie, où il avoit été envoyé pour faire des Observations astronomiques & physiques. 3 volumes. Amsterdam: Jean Catuffe, 1741.

[Format: octavo. Abridged French translation based on the German text of 1719 with the illustrations from the Dutch edition of 1727. A second and third impression of volumes 2 and 3 of the French translation appeared in 1742 and 1743—i.e. the parts about the Khoikhoi and the Dutch colony.]

[5] *The second, abridged German edition, based on the French translation*

Beschreibung des Vorgebürges der Guten Hoffnung, und derer darauf wohnenden Hottentotten. Worinnen von der natürlichen Beschaffenheit des Landes, von den Gebräuchen der Einwohner, ingleichen von der Einrichtung dasiger Holländischer Colonien zuverlässige Nachrich gegeben wird. Gezogen aus den Anmerckungen des Herrn M. Peter Kolbens, welche er während seines zehenjärrigen Aufenthaltes in dasigem Lande Verfertiget hat. Franckfurt/Leipzig, Peter Conrad Monath, 1745.

[Format: quarto. Not quite faithful German translation of the abridged French 1741 translation, with illustrations from *Capvt Bonae Spei Hodiernvm*.]

ANNEX TWO

STRUCTURE OF THE *NIEUWSTE EN BEKNOPTE BESCHRYVING VAN DE KAAP DER GOEDE-HOPE*

Voorbericht

[I; "Eerste stuk"] Beknopte beschryving van de Kaap der Goede Hope [numbered separately, pp. 1–140].

[II; Title page of the "Tweede stuk"] Dag-verhaal gehouden op een landtogt, door het land der kleine en groote Namaquas, Op bevel van den Gouverneur Door een Gezelschap van lxxxv Persoonen, onder Commando van den Capitain Hendrik Hop. Om binnenwaards ten Noorden van de Kaap der Goede-hope, liggende Landen nader te ontdekken [numbered separately, with its own title page, pp. 1–108].

[II, 1] Dag-verhaal gehouden op een landtogt, Door het land der kleine en groote Namaquas, Onder het Commando van den Capitein Hendrik Hop, om binnenwaarts ten Noorden van Cabo de Goede-hoop liggende Land, nader te ontdekken. Van den 16 July 1761, tot den 27 April 1762. [Compiled by C.F. Brink, pp. 1–79.]

[II, 2] Aan den wel edelen gestrengen heer, Ryk Tulbagh. [Compiled by Tielman Roos and Petrus Marais, "Cabo de Goede-hope den 18 Augustus 1762", pp. 80–91.]

[II, 3] Eerbiedig berigt gedaan aan den wel edelen gestrengen heer Ryk Tulagh [...] Behelzende de gesteldheid van het Gebergte, en de daarin gevonden Ertzen, dewelke op de jongstgedaanen Togt door het Land der *Namacquas* zyn ontdekt geworden. [Compiled by Carel Christoffel Rykvoet, pp. 92–95.]

[II, 4] Nieuwe ontdekking van den waschboom, Aan de Kaap der Goede-hope. Nevens eene Beschryving van deezen Boom in andere gewesten [pp. 96–108].

SOURCES AND BIBLIOGRAPHY

Archival sources

Amsterdam, Rijksprentenkabinet
Gordon Atlas (GA)

Den Haag, Nationaal Archief (NA)
Fagel Archief, 2515, 2533
Verzameling Buitenlandse Kaarten Leupe, 4.VEL

Houghton (Johannesburg), Brenthurst Library
Gordon Papers, MS107

Cape Town, Cape Archives (CA)
Bataafse Republiek (BR), BR377, BR549–550
Kaarten (Maps), M4
Microfilms, ZD/L4
Politieke Raad
Bylaes by Inkomende Briewe, C744, C751
Inkomende Briewe, C405, C416, C433, C733
Memories en Instruksies, C2341
Resolusies, C85, C139–140
Uitgaande Briewe, C1467–1468
Verklarings, C2440
Verbatim Copies (VC), VC75, VC592–597
Weeskamer (MOOC), MOOC 7/1/40

St. Maarten (Noord-Holland), Swellengrebel-Archief
Tekeningen, F1

Bibliography

[Mist, A. de]/Anonymous (1821). *Rélation d'un voyage en Afrique et en Amérique par Madame.* Namur: D. Gérard.

A South African Bibliography to the year 1925. (1979). 4 vols. London: Mansell.

Adams, P.G. (1962). *Travellers and travel liars 1660–1800.* Berkeley: University of California Press.

Adas, M. (1989). *Machines as the measure of men. Science, technology, and ideologies of Western dominance.* Ithaca: Cornell University Press.

Alberti, L. (1810a). *De Kaffers aan de Zuidkust van Afrika. Natuur- en Geschiedkundig beschreven.* Amsterdam: Maaskamp.

—— (1810b). *Zuid-Afrikaansche Gezichten.* Amsterdam: Maaskamp.

—— (1815). *Die Kaffern auf der Südküste von Afrika nach ihren Sitten und Gebräuchen aus eigener Ansicht beschrieben.* Gotha: Becker.

—— (1968). *Ludwig Alberti's account of the tribal life & customs of the Xhosa in 1807.* Cape Town: Balkema.

Alpers, S. (1989). *The art of describing. Dutch art in the seventeenth century.* Harmondsworth: Penguin.

Anderson, H.J. & Lloyd, A.C.G. (sa). *South Africa a century ago*. Cape Town/London: Maskew Miller/Blackwell.
"Anzeige neuer Schriften und Auszüge." (1791). *Magazin für das Neueste aus der Physik und Naturgeschichte* 7, nr. 2, pp. 100–131.
Axelson, E. (1973). *Portuguese in South-East Africa, 1488–1600*. Cape Town: Struik.
Barnard, A. (1992). *Hunters and herders of Southern Africa. A comparative ethnography of the Khoisan peoples*. Cambridge: Cambridge University Press.
Barnard, A.L. (1973). *The letters of Lady Anne Barnard to Henry Dundas, from the Cape and elsewhere, 1793–1803, together with her journal of a tour into the interior, and certain other letters*. (Ed. A.M. Lewin Robinson). Cape Town: Balkema.
Barnard, C.J. (1950). "Robert Jacob Gordon se loopbaan aan die Kaap." In *Argiefjaarboek* 13, pp. 319–446.
Barrell, J. (1988). "Being is perceiving. James Thomson and John Clare." In J. Barrell (Ed.), *Poetry, language and politics*, pp. 100–136. Manchester: Manchester University Press.
Barrow, J. (1801). *An account of travels into the interior of Southern Africa, in the years 1797 and 1798: including cursory observations on the geology and geography of the southern part of that continent; the natural history of such objects as occurred in the animal, vegetable, and mineral kingdoms, and sketches of the physical and moral characters of the various tribes of inhabitants surrounding the settlement of the Cape of Good Hope*. Vol. 1. London: Cadell & Davies.
—— (1801). *Voyage dans la partie méridionale de l'Afrique; fait dans les années 1797 et 1798*. Paris: Dentu. (Traduit de l'anglais par L. Degrandpré).
—— (1804). *An account of travels into the interior of southern Africa; in which is considered the importance of the Cape of Good hope to the different European powers, as a naval and military station; as a point of security to our Indian trade and settlements during a war and as a territorial acquisition and commercial emporium in time of peace* [...]. Vol. 2. London: Cadell & Davies.
—— (1806). *Travels into the interior of Southern Africa in which are described the character and the condition of the Dutch colonists of the Cape of Good Hope, and of the several tribes of natives beyond its limits: the natural history of such subjects as occurred in the animal, mineral, and vegetable kingdoms, and the geography of the southern extremity of Africa: comprehending also a topographical and statistical sketch of the Cape Colony*. 2 vols. London: Cadell & Davies.
—— (1847). *An autobiographical memoir of Sir John Barrow*. London: Murray.
Bartelings, N. (2001). "'Hier toont voor 't oog het tijtelblad den inhout in het boek vervat'. De rol van de titelprent in Frans van Mieris' Histori der Nederlandsche vorsten." In K. Bostoen, E. Kolfin & P.J. Smith (Eds.), *'Tweelinge eener dragt'. Woord en beeld in de Nederlanden (1500–1750)*, pp. 243–260. Hilversum: Verloren.
Barthes, R. (1968). "L'effet de réel", *Communications 11*, pp. 84–89.
Basalla, G. (1967). "The Spread of Western Science", *Science* 156, pp. 611–622.
Bassett, T.J. & Porter, P.W. (1991). "'From the best authorities': the mountains of Kong in the cartography of West Africa." *Journal of African History* 32, pp. 367–413.
Batten, C.L. (1978). *Pleasurable instruction. Form and convention in eighteenth-century travel literature*. Berkeley: University of California Press.
Becker-Moelands, M.A. (1985). *De juridische titelprent in de zeventiende eeuw*. Muiderberg: Coutinho.
Beer, R.R. (1977). *Unicorn: myth and reality*. New York: James J. Kerry. (*Einhorn: Fabelwelt und Wirklichkeit*).
Beinart, W. (1998). Men, science and nature in the eighteenth century and nineteenth century Cape. *Journal of Southern African Studies* 24, pp. 775–800.
Bergh, J.S. & Visagie, J.C. (1985). *The Eastern Cape frontier zone 1660–1980. A cartographic guide for historical research*. Durban: Butterworth.
Bergvelt, E. & Kistemaker, R. (Eds.) (1992). *De Wereld binnen handbereik: Nederlandse kunst- en rariteitenverzamelingen, 1585–1735*. Zwolle: Waanders.
Berman, R.A. (1998). *Enlightenment or empire. Colonial discourse in German culture*. Lincoln: University of Nebraska Press.

Bhila, H.H.K. (1982). *Trade and politics in a Shona kingdom. The Manyika and their African and Portuguese neighbours, 1575–1902.* Harlow (Essex): Longman.

Bierbrodt, J. (2000). *Naturwissenschaft und Ästhetik 1750–1810.* Würzburg: Köningshausen & Neuman.

Bindman, D. (2002). *Ape to Apollo. Aesthetics and the Idea of Race in the 18th Century.* London: Reaktion.

Blanckaert, C. (1988). "On the origins of French ethnology." In G.W. Stocking (Ed.), *Bones, bodies, behaviour. Essays on biological anthropology*, pp. 18–55. Madison: University of Wisconsin Press.

Blanke, H.W. (1997). *Politische Herrschaft und soziale Ungleichheit im Spiegel des Anderen. Untersuchungen zu den deutschsprachigen Reisebeschreibungen vornehmlich im Zeitalter der Aufklärung.* Waltrop: Spenner.

Blanton, C. (2002). *Travel writing: the self and the world.* New York: Routledge.

Blommaert, W. (1926). "'n Voorloper van Edw. GibbonWakefield." In *Mélanges d' histoire offerts à Henri Pirenne. Tome Premier*, pp. 43–54. Bruxelles: Vromant.

Blussé, L. & Ooms, I. (2002). "Kennis en Compagnie." In L. Blussé & I. Ooms (Eds.), *Kennis en Compagnie. De Vereenigde Oost-Indische Compagnie en de moderne Wetenschap*, pp. 7–11. Amsterdam: Balans.

Boemus, J. (1520). *Omnium gentium mores leges en ritus.* Augustae Vindelicorum: S. Grimm & M. Wirsung.

Boisacq, M.-J. (1993). "Le mythe du bon sauvage Hottentot", *Literator* 14, nr. 2, pp. 117–131.

Bokhorst, M. (1973). "'n Kunshistoriese waardering van die Le Vaillant-versameling." In J.C. Quinton (Ed.), *François le Vaillant in Suid-Afrika en sy versameling van 165 akwarelle 1781–1784.* Vol. 2, pp. 103–127. Kaapstad: Parlementsbiblioteek.

Bosman, W. (1704). *Nauwkeurige beschryving van de Guinese goud-, tand-, en slave kust.* Utrecht: Schouten.

Boucher, M. & Penn, N. (1992). *Britain at the Cape 1795 to 1803.* Houghton: Brenthurst Press.

Bourguet, M-N. (1996). "L'Explorateur." In M. Vovelle (Ed.), *L'Homme des Lumières*, pp. 289–293. Paris: Editions du Seuil.

—— (1997). "La collecte du monde: voyage et histoire naturelle (fin XVIIème siècle–début XIXème siècle)." In C. Blanckaert, P. Corsi, & J.C. Fisher (Eds.), *Le muséum au premier siècle de son histoire*, pp. 163–196. Paris: Éditions du museum.

Bourguet, M.-N., Licoppe, C. & Sibum, H.O. (Eds.) (2002). *Instruments, travel and science. Itineraries of precision from the seventeenth to the twentieth century.* London: Routledge.

Böving, J.G. (1712). *Herrn Joh. Georg Bövings/königlichen dänischenMissionarii zu Tranquebar in Ost-Indien/Curieuse Beschreibung und Nachricht von den Khoikhoi.* sl: sa.

Boyle, R. (1666). "General Heads for a Natural History of the Countrey, Great or Small." *Philosophical Transactions* 1, nr. 10, pp. 186–189.

Bradlow, E. & Bradlow, F. (Eds.) (1979). *William Sommerville's narrative of his journey to the eastern Cape frontier and to Lattakoe 1799–1802.* Cape Town: Van Riebeeck Society.

Broberg, G. (1983). "*Homo sapiens.* Linnaeus's classification of man." In T. Frängsmyr (Ed.), *Linnaeus. The man and his work*, pp. 156–194. Berkeley: University of California Press.

Bucquoi, Jakob de (1744). *Aanmerkelyke ontmoetingen in de zestien jaarige reize naa de Indien.* Haarlem: Bosch.

Buffon, G.-L. de (1749). *Histoire naturelle générale et particulière. Tome premier.* Paris: Impremerie royale.

Bulwer, J. (1974). *Chirologia or the natural language of the hand and Chironomia or the art of manual rhetoric.* Carbondale: Southern Illinois University Press.

Bunn, D. (1994). "Our wattled cot. Mercantile and domestic space in Thomas Pringle's African landscapes." In W.J.T. Mitchell (Ed.), *Landscape and power*, pp. 127–174. Chicago: University of Chicago Press.

Burke, P. (2000). *A social history of knowledge. Gutenberg to Diderot.* Cambridge: Polity Press.

Cahn, M. (1997). "Die Rhetorik der Wissenschaft im Medium der Typographie. Zum Beispiel die Fußnote." In H.-J. Rheinberger, M. Hagner & B. Wahrig-Schmidt (Eds.), *Räume des Wissens. Repräsentation, Codierung, Spur*, pp. 91–110. Berlin: Akademie Verlag.

Callon, M. (1986). "Some elements of a sociology of translation; domestication of the scallops and the fisherman of St Brieuc Bay." In J. Law (Ed.), *Power, action and belief. A new sociology of knowledge?*, pp. 196–233. London etc.: Routledge/Kegan Paul.

Cameron, J.M.R., McLaren, G. & Cooper, W. (1999). "Bushmanship: the explorers' silent partner." *Australian Geographer* 30, nr. 3, pp. 337–353.

Campbell, J. (1974/1815). *Travels in South Africa.* Cape Town: Struik.

Camper, P. (1786). "Zusatz zu der vorhergehenden Abhandlung aus einem Schreiben an die Naturforschende Gesellschaft." *Schriften der Berlinischen Gesellschaft naturforschender Freunde* 7, nr. 2, pp. 219–226.

—— (1791). *Verhandeling over het natuurlijk verschil der wezenstrekken in menschen van onderscheiden landaart en ouderdom* [...]. Utrecht: Wild & Altheer.

Carey, D. (1997). "Compiling nature's history. Travellers and travel narratives in the early Royal Society." *Annals of Science* 54, pp. 269–292.

Chappey, J.-L. (2002). *La Société des Observateurs de l'Homme (1799–1804). Des anthropologues au temps de Bonaparte.* Paris: Société des études robespierristes.

Charlton, D.G. (1984). *New images of the natural in France. A study in European cultural history 1750–1800.* Cambridge: Cambridge University Press.

Chidester, D. (1996). *Savage systems. Colonialism and comparative religion in Southern Africa.* Cape Town: University of Cape Town Press.

Cicero, M.T. (1949). *De Inventione. De optimo genere oratorum. Topica.* London/Cambridge: Heinemann/Harvard University Press.

Clifford, J. (1988). *The predicament of culture. Twentieth-century ethnography, literature, and art.* Cambridge: Harvard University Press.

Coetzee, A. (1998). "Duytsman altyd kallum: Icke Hottentots dootmakom." *Alternation* 5, nr. 2, pp. 36–42.

Coetzee, C. (1995). "Two versions of a journey into the interior." *Pretexts* 5, nrs. 1–2, pp. 66–83.

Coetzee, C.G. (1948). "Die Kompanjie se besetting van Delagoabaai." *Argiefjaarboek* 11, nr. 2, pp. 171–276.

—— (1954). *Die stryd om Delagoabaai en die suidooskus, 1600–1800.* Dissertation, Universiteit van Stellenbosch.

Coetzee, J.M. (1988). *White writing. On the culture of letters in South Africa.* New Haven: Yale University Press.

Contarini, S. (2004). "François Levaillant: un voyageur 'éclairé' entre histoire naturelle et anthropologie." In P. Despoix & J. Fetscher (Eds.), *Cross-cultural encounters and constructions of knowledge in the 18th and 19th century. Non-European and European travel of exploration in comparative perspective*, pp. 259–282. Kassel: Kassel University Press.

Copans, J. & Jamin, J. (1978). *Aux origines de l'anthropologie française. Les Mémoires de la Société des Observateurs de l'Homme en l'an VIII.* Paris: Le Sycomore.

Corbett, M. & Lightbown, R. (1979). *The comely frontispiece. The emblematic title-page in England, 1550–1660.* London: Routledge & Paul.

Couzens, T. (2004). *Battles of South Africa.* Claremond: David Philip.

Cro, S. (1990). *The noble savage. Allegory of freedom.* Waterloo: Wilfred Laurier University Press.

Croix, A.P. de la (1705). *Algemeene wereld-beschrijving.* Amsterdam: Halma.

Cullinan, P. (1992). *Robert Jacob Gordon 1743–1795. The man and his travels at the Cape.* Cape Town: Struik.

"Dag-Verhaal eener Reize naar de binnenlanden van Afrika beoosten de Kaap de Goede Hoop geleegen, in den jaare 1803 gedaan door Z.E. den Gouverneur en Generaal en Chef J.W. Janssens verzeld door deszelfs Aide de Camp Capt. Paravicini di Capelli, den 1n Chirurgyn by het Escadron Ligte Dragonders Passet, den Heer Dirk G. van Reenen en deszelfs zoon Daniel van Reenen." (1937). In W. Blommaert & J.A. Wiid (Eds.), *Die joernaal Dirk Gysbert van Reenen 1803*. Kaapstad: Van Riebeeck Vereniging.

Daston, L. (1988). "The factual sensibility." *Isis* 79, pp. 452–470.

—— (1991). "Marvelous facts and miraculous evidence in early modern Europe." *Critical Inquiry* 18, pp. 93–124.

Daston, L. & Park, K. (2001). *Wonders and the order of nature, 1150–1750*. New York: Zone Books.

De Foigny, G. (1676). *Les Avantures de Jacques Sadeur dans la decouverte et le voyage de la Terre Australe*. Genève: C. Barbin.

De Gérando, J.-M. (1978) [1800]. *Considérations sur les diverses méthodes a suivre dans l'observation des peuples sauvages*. In J. Copans & J. Jamin (Éds.), *Aux origines de l'anthropologie française. Les Mémoires de la Société des Observateurs de l'Homme en l'an VIII*, pp. 126–169. Paris: Le Sycomore.

De Groot, E. (2005). "The Earliest Eyewitness Depictions of Khoikhoi." *Itinerario* 29, nr. 1, pp. 17–50.

De la Caille, N. (1763). *Journal historique du voyage fait au Cap de Bonne-Esperance par feu M. l'Abbé de la Caille précédé d'un discours sur la vie de l'auteur, suivi de remarques & de réflexions sur les coutûmes des Hottentots & des habitans du Cap*. Paris, Guillyn.

De Lahontan, L.A. (1703). *Nouveaux Voyages*. La Haye: l'Honoré.

Dear, P. (1985). "Totius in verba. Rhetoric and authority in the early Royal Society." *Isis* 76, pp. 145–161.

—— (2001). *Revolutionizing the sciences. European knowledge and its ambitions, 1500–1700*. Princeton: Princeton University Press.

Demel, W. (1992). "Wie die Chinesen gelb wurden. Ein Beitrag zur Frühgeschichte der Rassentheorien." *Historische Zeitschrift* 255, pp. 69–93.

De Wet, G.C. (Ed.) (1981). *Resolusies van die Politieke Raad. Deel IX, 1735–1739*. Pretoria: Staatsdrukker.

Donat, D. (1966). "Zu Buchtiteln und Titelblättern der Barockzeit." In D. Gerhardt, W. Weintraub & H.-J. zum Winkel (Eds.), *Orbis Scriptus. Dmitrij Tschilewskij zum 70. Geburtstag*, pp. 163–173. München: Wilhelm Finck.

Dos Santos Lopes, M. (1992). *Afrika. Eine neue Welt in deutschen Schriften des 16. und 17. Jahrhunderts*. Stuttgart: Franz Steiner Verlag.

Douglas, B. (1999). "Art as ethno-historical text: science, representation and indigenous presence in eighteenth and nineteenth century oceanic voyage literature." In N. Thomas & D. Losche (Eds.), *Double Vision. Art Histories and Colonial Histories in the Pacific*, pp. 65–102. Cambridge: Cambridge University Press.

—— (2003). "Seaborne ethnography and natural history of man." *The Journal of Pacific History* 38, nr. 1, pp. 3–27.

Drayton, R. (2000). *Nature's government. Science, imperial Britain, and the 'Improvement' of the world*. New Haven: Yale University Press.

Duchet, M. (1985). *Le partage des savoirs. Discours historique et discours ethnologique*. Paris: Éditions la Découverte.

Dyck, J. (1969). *Ticht-Kunst. Deutsche Barockpoetik und Rhetorische Tradition*. Bad Homburg: Gehlen.

Dyer, R.A. (sa). "Colonel Robert Jacob Gordon's contribution to South African botany." *South African Biological Society Pamphlets* 14, pp. 44–62.

Eddy, J.H. (1994). "Buffon's *Histoire naturelle* history? A critique of recent interpretations." *Isis* 85, pp. 644–661.

Edney, M.H. (1999). "Reconsidering enlightenment geography and mapmaking: Reconnaissance, mapping, archive." In D.N. Livingstone & C.W. J. Withers (Eds.), *Geography and enlightenment*, pp. 165–198. Chicago: University of Chicago Press.

Eilberg-Schwartz, H. (1990). *The savage in Judaism. An anthropology of Israelite religion and ancient Judaism*. Bloomington: Indiana University Press.

Ellen, F.L. (1976). "The development of anthropology and colonial policy in the Netherlands: 1800–1960." *Journal of the History of Behavioral Sciences* 12, nr. 4, pp. 303–324.

Elphick, R. (1985). *Khoikhoi and the founding of white South Africa.* Johannesburg: Ravan Press.

Engelbrecht, S.P. (sa). *Die Kaapse predikante van die sewentiende en agtiende eeu.* Kaapstad/ Pretoria: H.A.U.M./J.H. de Bussy.

Enklaar, I.H. (1988). *Life and work of dr. J. Th. van der Kemp, 1747–1811. Missionary pioneer and protagonist of racial equality in South Africa.* Cape Town: Balkema.

Evenhuis, R.B. (1974). *Ook dat was Amsterdam. Deel IV. De kerk der hervorming in de achttiende eeuw: de grote crisis.* Baarn: Ten Have.

Fabian, J. (1983). *Time and the Other. How anthropology makes its object.* New York: Columbia University Press.

—— (1990). "Presence and representation. The other and anthropological writing." *Critical Inquiry 16*, pp. 753–772.

—— (2000). *Out of our minds. Reason and madness in the discovery of central Africa.* Berkeley: University of California Press.

Faidutti, B. (1996). *Images et connaissance de la licorne (Fin du Moyen-Age—XIX*ème *siècle).* Dissertation Université Paris XII, available on the internet: www.faidutti.free.fr/licornes/these.

Fauvelle-Aymar, F-X. (2002). *L'invention du Hottentot. Histoire du regard occidental sur les Koisan (XV*e*–XIX*e *siècle).* Paris: Publications de la Sorbonne.

Firla, M. (1994). "Kants Bild von den Khoi-Khoin (Südafrika)." *Tribus. Jahrbuch des Linden Museums* 43, pp. 60–94.

Fleck, L. (1979). *Genesis and development of a scientific fact.* Chicago: University of Chicago Press.

Forbes, V.S. (1951–1952). "Further notes on colonel Gordon." *Africana News and Notes* 9, pp. 84–94.

—— (1965). *Pioneer travellers of South Africa. A geographical commentary upon routes, records, observations and opinions of travellers at the Cape, 1750–1800.* Cape Town: Balkema.

—— (1973). "Le Vaillant's travels in South Africa, 1781–4." In *F. le Vaillant. Traveller in South Africa and his collection of 165 water-colour paintings 1781–1784*, pp. 31–110. Cape Town: Parliament Library.

Forster, G. (1784). "Vorrede des Herausgebers." In A. Sparmann [Sparrman], *Reise nach dem Vorgebirge der guten Hoffnung.* Berlin: Haude & Spener.

—— (1967–1970). *Werke in vier Bänden.* Frankfurt am Main: Insel.

—— (1992). *Georg Forsters Werke. Sämtliche Schriften, Tagebücher, Briefe. Rezensionen. Elfter Band.* Berlin: Akademie Verlag.

Foster, R. (2004). "Die boekstawing van die sewentiende-eeuse landreise. Die omsetting van gebeurtenis tot feit." *Tydskrif vir Nederlands en Afrikaans* 11, pp. 20–40.

Foucault, M. (2003). *The Birth of the Clinic: An Archaeology of Medical Perception.* London: Routledge. (Translation of *Naissance de la clinique: une archéologie du regard médical.*)

Francken, J. (1761). *Rampspoedige reize van het O.I. schip De Naarstigheid, in de terugreize van Batavia over Bengale naar Holland, waar van de schepelingen zyn t'huis gekomen den 13den Juny 1760, met 't schip Scholtenburg.* Haarlem: Bosch.

Freeman, M. (1976). *The unicorn tapestries.* New York: The Metropolitan Museum of Art.

Freund, W.M. (1971). *Society and government in Dutch South Africa. The Cape and the Batavians, 1803–6.* Dissertation, Yale University.

Freundlieb, D. (1995). "Foucault's theory of discourse and human agency." In C. Jones & R. Porter (Eds.), *Reassessing Foucault. Power, medicine and the body*, pp. 152–180. London: Routledge.

Fryer, P. (1984). *Staying power: the history of black people in Britain.* London: Pluto Press.

Fuchs, T. (2006). The European China-receptions from Leibnitz to Kant. *Journal of Chinese Philosophy* 33, pp. 35–49.

Fulford, T., D. Lee & P.J. Kitson. (2004). *Literature, science and exploration in the Romantic era: bodies of knowledge.* Cambridge: Cambridge University Press.

Gandelmann, C. (1986). *Le regard dans le texte. Image et écriture du Quattrocento au XXe siècle.* Paris: Méridiens Klincksieck.

Gebhard jr., J.F. (1881–1882). *Het leven van Mr. Nicolaas Cornelisz Witsen (1641–1717)* 2 vols. Utrecht: Leeflang.

Gelderblom, A.J. (1997). "Schrijvende leidsvrouw en kindervriendin: Anna Barbara van Meerten-Schilperoort (1778–1853)." *Nederlandse Letterkunde* 2, pp. 29–44.

Gerritsen, W.P. (2003). *De eenhoorn en de geleerden. Het debat over het bestaan van de eenhoorn van de zestiende tot de negentiende eeuw.* Leiden: Universiteitsbibliotheek Leiden.

Giliomee, H. (1981). "Processes in the development of the Southern African Frontier." In H. Lamar & L.Thompson (Eds.), *The Frontier in history. North America and Southern Africa compared*, pp. 76–119. New Haven: Yale University Press.

—— (1989). "The Eastern Frontier, 1770–1812." In R. Elphick & H. Giliomee (Eds.), *The Shaping of South African Society, 1652–1840*, pp. 421–71. Cape Town: Maskew Miller/Longman.

Gitelman, L.L. (1991). *The world recounted; science and narrative in early nineteenth-century exploration accounts.* Unpublished dissertation, Columbia University.

Glenn, I. (1997). "Le Vaillant's Voyages and the circumstances surrounding the publication of the Voyages." In P.E. Westra & L.T. Jones (Eds.), *The love of books*, pp. 85–93. Cape Town: South African Library.

Godée Molsbergen, E.C. (Ed.) (1916–1932). *Reizen in Zuid-Afrika in de Hollandse tijd. Vierde deel. Tochten in het Kafferland 1776–1805.* 4 vols. 's-Gravenhage: Nijhoff.

Godlewska, A. (1995). "Map, text and image. The mentality of enlightened conquerors. A new look at the *Description de l'Egypte.*" *Transactions of the Institute of British Geographers* 20, pp. 5–28.

—— (1999). *Geography unbound. French geographic science from Cassini to Humboldt.* Chicago: University of Chicago Press.

Grafton, A. (1992). *New worlds, ancient texts. The power of tradition and the shock of discovery.* Boston: Belknap Press.

Grant, C.H.B. (1957). "Levaillant's travels in South Africa. 1781–1784." *The Ostrich* 28, pp. 83–97.

Gritsch, E.W. (1994). "The Jews in Reformation Theology." In M. Perry & F.M. Schweitzer (Eds.), *Jewish-Christian encounters over the centuries. Symbiosis, prejudice, holocaust, dialogue*, pp. 197–214. New York: Peter Lang.

Guelke, L. & Guelke, J.L. (2004). "Imperial eyes on South Africa. Reassessing travel narratives." *Journal of Historical Geography* 30, pp. 11–31.

Gulliver, P.H. (1985). "An Applied Anthropologist in East Africa during the Colonial Era." In R. Grillo en A. Rew (Eds.), *Social Anthropology and Development Policy*, pp. 37–57. London/New York: Tavistock.

Haafner, J. (1820). *Lotgevallen en vroegere zeereizen van Jacob Haafner volgens deszelfs nagelatene papieren uitgegeven door C.M. Haafner.* Amsterdam: Johannes van der Heij.

—— (1992). *De werken van Jacob Haafner.* Vol. 1. (J.A. de Moor & P.G.E.I.J. van der Velde (Eds.)). Zutphen: Walburg Pers.

—— (1993). *Verhandeling over het nut der zendelingen en zendelings-genootschappen.* (J.A. de Moor & P.G.E.I.J. van der Velde (Eds.)). Hilversum: Verloren.

Hacking, I. (1982). "Language, truth and reason." In M. Hollis & S. Lukes (Eds.), *Rationality and relativism*, pp. 48–66. Oxford: Basil Blackwell.

Hall, M. (2000). *Archeology and themodernworld. Colonial transcripts in South Africa and the Chesapeake.* London: Routledge.

Hallema, A. (1932). "Een bezoek van Mr. Hendrik Swellengrebel aan den Kafferkapitein Jeramba." *Zuid-Afrika* 9, nr. 9, pp. 131–137.

—— (1951). *Die Kaap in 1776–1777. Akwarelle van Johannes Schumacher uit die Swellengrebel-argief te Breda.* 's-Gravenhage: Stols.

Hallowell, A.I. (1960). "The beginnings of anthropology in America." In A. Irving Hallowell (Ed.), *Selected papers from the American Anthropologist 1888–1920*, pp. 1–90. Evanston: Row, Peterson & Company.

Halttunen, K. (1995). "Humanitarianism and the pornography of pain in Anglo-American culture." *American Historical Review 100*, nr. 2, pp. 303–334.

Harbsmeier, M. (1994). *Wilde Völkerkunde. Andere Welten in deutschen Reiseberichten der frühen Neuzeit.* Frankfurt: Campus.

—— (1995). "Towards a prehistory of ethnography: Early modern German travel writing as traditions of knowledge." In H.F. Vermeulen & A.A. Roldán (Eds.), *Fieldwork and footnotes. Studies in the history of European anthropology*, pp. 19–38. London: Routledge.

Harley, J.B. (2001), *The new nature of maps. Essays in the history of cartography*. Baltimore: The John Hopkins University Press.

Harris, S.J. (1998) "Long-distance Corporations, Big Sciences, and the Geography of Knowledge." *Configurations* 6, pp. 269–304.

Hartog, F. (1980). *Le miroir d'Hérodote. Essai sur la représenation de l'autre.* Paris: Gallimard.

Haskell, F. & Penny, N. (1982). *Taste and the antique. The lure of classical sculpture 1500–1900.* New Haven: Yale University Press.

Heawood, E. (1969). *History of geographical discovery in the seventeenth and eighteenth centuries.* New York: Octagon.

Heese, H.F. (2005). *Groep sonder grense. Die rol en status van die gemengde bevolking aan die Kaap, 1652–1795.* Pretoria: Protea.

Hentschel, U. (1999). *Studien zur Reiseliteratur am Ausgang des 18. Jahrhunderts.* Frankfurt am Main: Peter Lang.

Hodgen, M.T. (1964). *Early anthropology in the sixteenth and seventeenth centuries.* Philadelphia: University of Pennsylvania Press.

Hoogvliet, M. (1999). *Mappae mundi: scriptura et pictura. Textes, images et herméneutique des mappemondes du Moyen Age long (XIIIe–XVIe siècles).* Dissertation, Rijksuniversiteit Groningen.

Horowitz, E. (1992). "'A different mode of civility': Lancelot Addison on the Jews of Barbary." In D. Wood (Ed.), *Christianity and Judaism*, pp. 309–326. Oxford: Blackwell.

Hudson, N. (2004). "'Hottentots' and the evolution of European racism." *Journal of European Studies* 34, pp. 308–332.

Huigen, S. (1996a). *De weg naar Monomotapa: Nederlandstalige representaties van geografische, historische en sociale werkelijkheden in Zuid-Afrika.* Amsterdam: Amsterdam University Press.

—— (1996b). "Travellers to Monomotapa. The representation of Southern Africa by the Dutch in the seventeenth century." *History and anthropology* 9, pp. 207–230.

—— (1998). "Natural history and the representation of South Africa in the eighteenth century." *Journal of Literary Studies 14*, pp. 67–80.

—— (2001). "Expedities vanuit Fort Lijdzaamheijd: representaties van het binnenland van zuidelijk Afrika door voc-reizigers in de achttiende eeuw." *Tydskrif vir Nederlands en Afrikaans* 8, pp. 35–57.

Hulme, P. (1986). *Colonial encounters. Europe and the native Caribbean 1492–1797.* London: Methuen.

Jacobs, E.M. (2000). *Koopman in Azië. De handel van de Verenigde Oost-Indische Compagnie tijdens de 18de eeuw.* Zutphen: Walburg Pers.

Jansen, E. (1997). "'Wij meenden evenwel voor Zuid-Afrika onzen plicht gedaan te hebben.' Die Suid-Afrikaanse gedeelte van dr. Aletta Jacobs van Amsterdam se *Reisbrieven uit Afrika en Azië* (1913)." *Tydskrif vir Nederlands en Afrikaans* 4, pp. 1–27.

Jeffreys, K.M. (Ed.) (1928). *Kaapse archiefstukken lopende over het jaar 1780* Kaapstad: Cape Times Beperkt.

Johnson, D. (1998). "De Mist, race and nation." *Alternation* 5, nr. 1, pp. 85–97.

Joppien, R. & B. Smith (1985–1988). *The art of Captain Cook's voyages*. 4 vols. New Haven/London: Yale University Press.

"Journaal en verhaal eener landreyse in den jaare 1803 door den gouverneur en generaal deeser colonie J.W. Janssens door de binnenlanden van Zuid-Africa gedaan." (1932). In E.C. Godée Molsbergen (Ed.), *Reizen in Zuid-Afrika in de Hollandse tijd. Vierde deel. Tochten in het Kafferland 1776–1805*, pp. 100–209. 's-Gravenhage: Nijhoff.

Katz, D.S. (1992). "The Phenomenon of Philo-Semitism." In D. Wood (Ed.), *Christianity and Judaism*, pp. 327–362. Cambridge: Blackwell.

Keegan, T. (1996). *Colonial South Africa and the origins of the racial order*. Cape Town: David Philip.

Kemp, J.T. van der (1804). "An account of the religion, customs, population, government, language, history, and natural productions of Caffraria." In *Transactions of the London Missionary Society. Vol. 1. From its Institutions in the year 1795 to the end of the year 1802*, pp. 432–468. London: Williams.

Kerby, P.R. (1942). "The Heerenlogement and its visitors. The story of one of South Africa's 'Historical Monuments'." *South African Journal of Science 38*, pp. 352–386.

Kirshenblatt-Gimblett, B. (1991). "Objects of ethnography." In I. Karp & S.D. Lavine (Eds.), *Exhibiting cultures. The poetics and politics of museum display*, pp. 386–443. Washington/London: Smithsonian Institution Press.

"Klad journaal der reize door den commissaris generaal Mr. J.A. de Mist en deszelfs gevolg in de binnenlanden van Afrika." (1911). In G.M. Theal (Ed.), *Belangrijke historische dokumenten*, pp. 121–187. Cape Town: Van de Sandt de Villiers.

Kloek, J. & Tilmans, K. (Eds.) (2002). *Burger. Een geschiedenis van het begrip 'burger' in de Nederlanden van de Middeleeuwen tot de 21ste eeuw*. Amsterdam: Amsterdam University Press.

Kloek, J. & Mijnhardt, W. (2001). *1800: Blauwdrukken voor een samenleving*. Den Haag: Sdu.

Knebel, K.L. von (1802). "Bedenken über die Abbildung des Einhorns, in Hrn. Barrow's Reis nach dem Innern des südlichen Theiles von Afrika." *Der neue Teutsche Merkur* 1, pp. 94–100.

Knox, R. (1681). *An historical relation of the island Ceylon, in the East-Indies: together with an account of the detaining in captivity the author and divers other Englishmen now living there, and of the author's miraculous escape*. London: Richard Chiswell.

Koeman, C. (1952a). "Landmeters in dienst van de V.O.C. in de Kaapkolonie." *Tijdschrift voor kadaster en landmeetkunde* 68, pp. 93–99.

—— (1952b). *Tabvlae geographicae quibus Colonial Bonae Spei Antiqva depingitvr*. Kaapstad: H.A.U.M.

—— (1970). *Joan Blaeu and his Grand Atlas. Introduction to the facsimile edition of le grand atlas*. Amsterdam: Theatrum Orbis Terrarum.

Kolb, P. (1719). *Capvt Bonae Spei Hodiernvm: das ist, Vollstandige Beschreibung des Afrikanischen Vorgeburges der Guten Hofnung*. Nürnberg: Monath.

—— *The present state of the Cape of Good-Hope: or, A particular account of the several nations of the Hottentots: their religion, government, laws, customs, ceremonies, and opinions*. London: W. Innys.

—— (1731). *The present state of the Cape of Good-Hope: vol. II. Containing, the natural history of the Cape; or, A particular description of all the sorts of animals in that neighbourhood*. London: W. Innys.

—— (1741). *Description du Cap de Bonne-Esperance, où l'on trouve tout ce qui concerne l'histoire naturelle du pays*. 3 vols. Amsterdam: Jean Catuffe.

—— (1745). *Beschreibung des Vorgebürges der Guten Hoffnung, und derer darauf wohnenden Hottentotten*. Franckfurt: Peter Conrad Monath.

—— (1968/1731). *The present state of the Cape of Good Hope*. New York: Johnson Reprint Corporation.

Kolbe [Kolb], P. (1727). *Naaukeurige en uitvoerige beschryving van Kaap de Goede Hoop*. 2 vols. Amsterdam: Balthazar Lakeman.

Kommers, J. (1996). "Snouck Hurgronje als koloniaal etnograaf: *De Atjèhers* (1893–1894)." *Sharqiyyât* 8, nr. 2, pp. 87–115.

Koningsveld, H. (1982). *Het verschijnsel wetenschap. Een inleiding tot de wetenschapsfilosofie.* Meppel: Boom.

Konvitz, J. (1987). *Cartography in France, 1660–1848. Science, engineering, and statecraft.* Chicago: University of Chicago Press.

Koolhaas-Grosfeld, E. (2003). "Een nieuwe opdracht voor de kunst. Exotisme in beweging. Etnografische beelden rond 1800." In J. de Jong e.a. (Ed.), *Het exotische verbeeld 1550–1950. Boeren en verre volken in de Nederlandse kunst*, pp. 207–250. Zwolle: Waanders. (Nederlands Kunsthistorisch Jaarboek, 53).

Krynauw, D.W. (1999). *Beslissing by Blouberg. Triomf en tragedie van die stryd om die Kaap.* Kaapstad: Tafelberg.

Kuhn, T.S. (1970). *The Copernican revolution. Planetary astronomy in the development of Western thought.* Cambridge: Harvard University Press.

Künne, W. (2003). *Conceptions of Truth.* Oxford: Clarendon Press.

Kury, L. (1998). "Les instructions de voyage dans les expéditions scientifiques françaises (1750–1830)." *Revue d'Histoire des Sciences* 51, nr. 1, pp. 65–91.

—— (2001). *Histoire naturelle et voyages scientifiques (1780–1830).* Paris: L'Harmattan.

Lagueux, O. (2003). "Geoffroy's Giraffe: The Hagiography of a Charismatic Mammal." *Journal of the History of Biology* 36, pp. 225–247.

Lambert, D & Lester, A. (2004). "Geographies of colonial philanthropy." *Progress in Human Geography* 28, nr. 3, pp. 320–341.

Landau, P.S. (1996). "With camera and gun in Southern Africa. Inventing the image of Bushmen c. 1880 to 1935." In P. Skotnes (Ed.), *Miscast. Negotiating the presence of the Bushmen*, pp. 129–142. Cape Town: University of Cape Town Press.

Laqueur, T.W. (1989). "Bodies, details, and the humanitarian narrative." In L. Hunt (Ed.), *The new cultural history*, pp. 176–204. Berkeley: University of California Press.

Latour, B. (1983). "Give me a laboratory and I will raise the world." In K.D. Knorr-Cetina & M. Mulkay (Eds.), *Science observed. Perspectives on the social study of science*, pp. 141–170. London etc.: Sage.

—— (1987). *Science in action. How to follow scientists and engineers through society.* Cambridge (Mass.): Harvard University Press.

—— (1988). *The Pasteurization of France.* Cambridge (Mass.): Harvard University Press.

—— (1990). "Drawing things together." In M. Lynch & S. Woolgar (Eds.), *Representation in scientific practice*, pp. 19–68. Cambridge (Mass.): MIT Press.

—— (1996). *Aramis or the love of technology.* Cambridge (Mass.): Harvard University Press.

Latour, B. & Woolgar, S. (1986). *Laboratory life. The construction of scientific facts.* Princeton: Princeton University Press.

Lausberg, H. (1960). *Handbuch der literarischen Rhetorik.* München: Max Hüber Verlag.

Law, J. (1986). "On the methods of long-distance control: vessels, navigation and the Portuguese route to India." In J. Law (Ed.), *Power, action and belief. A new sociology of knowledge?*, pp. 234–263. London etc.: Routledge/Kegan Paul.

Le Vaillant, F. (1790). *Voyage de Monsieur le Vaillant dans l'interieur de l'Afrique, par le Cap de Bonne-Esperance, dans les annees 1780, 81, 82, 83, 84 & 85.* Paris: Leroy.

—— (1791–1798). *Reize in de binnenlanden van Afrika, langs de Kaap de Goede Hoop, in de jaren MDCCLXXX tot MDCCLXXXV gedaan door Le Vaillant.* (Translation of *Voyage* and *Second Voyage* by J.D. Pasteur). 5 vols. Leyden/Amsterdam: Honkoop/Allart.

—— (1795) [l'An IVe]. *Second Voyage de Monsieur Le Vaillant dans l'intérieur de l'Afrique, par le Cap de Bonne-Espérance, dans les années 1783, 84 et 85.* 2 vols. Paris: Chez H.J. Jansen.

—— (2007). *Travels into the Interior of Africa via the Cape of Good Hope. Volume 1.* Cape Town: Van Riebeeck Society.

Le Comte, L.D. (1696). *Nouveaux Mémoires sur l'état présent de la Chine.* Paris: J. Anisson.

Legassick, M. (1992). "The northern Frontier to c. 1840. The rise and decline of the Griqua people." In R. Elphick & H. Giliomee (Eds.), *The Shaping of South African Society, 1652–1840*, pp. 358–420. Cape Town: Maskew Miller Longman.

Lester, A. (2001). *Imperial networks. Creating identities in nineteenth-century South Africa and Britain*. London: Routledge.

Lestringant, F. (2003). *Sous la leçon des vents. Le monde d'André Thevet, cosmographe de la Renaissance*. Paris: Presses de l'Université de Paris-Sorbonne.

Leupe, A.P. (Ed.) (1852). "Instructie, waerna den capitein der vierde comp. burger dragonders aan Stellenbosch, Hendrik Hop, nevens de verdere persoonen, gaande ter ontdeckinge van 't binnenwaerts ten noorden van hier leggende land, sig gedurende desen weg zullen hebben te gedragen." *Kronijk van het Historisch Genootschap* 8, pp. 372–378.

Lichtenstein, H. (1811–1812). *Reisen im südlichen Africa in den Jahren 1803, 1804, 1805 und 1806*. 2 vols. Berlin: Salfeld.

—— (1973). *Foundation of the Cape. About the Bechuanas*. (Ed. O.H. Spohr). Cape Town: Balkema.

Licoppe, C. (1996). *La formation de la pratique scientifique. Le discours de l'expérience en France et en Angleterre (1630–1820)*. Paris: La Découverte.

Linnaeus, C. (1760). *Instructio Perigrinatoris*. In Caroli Linnaei, *Amoenitates Academicae, seu Dissertationes variae physicae, medicae, botanicae, antehac seorsi meditae*, pp. 298–313. Holmiae: Leurentii Salvii.

Livingstone, D. (2004). "Cultures of science." In J.S. Duncan, N.C. Johnson & R.H. Schein (Eds.), *A companion to cultural geography*, pp. 139–150. Malden: Blackwell.

Lloyd, C. (1970). *Mr. Barrow of the Admiralty. A life of Sir John Barrow 1764–1848*. London: Collins.

Lobo, G. (Jeronimo) (1728). *Relation historique d'Abissinie*. Paris: Veuve A.-U. Coustelier & J. Guérin.

—— (1984). *The itinerario of Jeronimo Lobo*. (Ed. D.M. Lockhart). London: Hakluyt Society.

Locke, J. (1997/1690). *An essay concerning human understanding*. London: Penguin.

Lovejoy, A.O. (1960). "The supposed primitivism of Rousseau's Discourse on Inequality." In A.O. Lovejoy, *Essays in the History of Ideas*, pp. 14–37. New York: Capricorn Books/G.P. Putnam's Sons.

—— (1965). *The great chain of being. A study of the history of an idea*. New York: Harper & Row.

Ludolf, H. (1681). *Historia Aethiopica, sive brevis et succinta descriptio regni Habessionorum, quod vulgo male Presbyteri Johannis vocatur: Libri quator*. Francofurti ad Moenum: Zunner.

Lux, D.S. en Cook, H.J. (1998). "Closed circles or open networks? Communicating at a distance during the scientific revolution." *History of Science* 36, pp. 179–211.

MacMillan, K. (2003). "Sovereignty 'More Plainly Described': Early English maps of North America, 1580–1625." *Journal of British Studies* 42, pp. 413–447.

Magli, P. (1989). "The face and the soul." In M. Feher (Ed.), *Fragments for a history of the human body*. Vol. 2, pp. 87–127. New York: Zone.

Maltby, W.S. (1971). *The Black Legend in England. The development of anti-Spanish sentiment, 1558–1660*. Durham: Duke University Press.

Marouby, C. (1990). *Utopie et primitivisme. Essai sur l'imaginaire anthropologique à l'âge classique*. Paris: Seuil.

Marshall, J. (1988). British assessments of the Dutch in Asia in the age of Raffles. *Itinerario 12*, nr. 1, pp. 1–13.

Mason, P. (1990). *Deconstructing America*. London: Routledge.

—— (1998). *Infelicities. Representations of the Exotic*. Baltimore: The Johns Hopkins University Press.

Masson, F. (1776). "An account of the three journeys from Cape Town into the southern parts of Africa." *Philosophical transactions of the Royal Society of London 66*, Vol. 1, pp. 268–317.

Mauersberger, G. (1994). "Der Gründer des Berliner Zoologischen Gartens, Martin Hinrich Lichtenstein (1780–1857). Eine biografisch Skizze." *Bongo, Berlin 23*, pp. 3–34.

Mazzolini, R.G. (2005). "L'interpretazione simbolica della pigmentazione umana nell'antropologia fisica del primo Ottocento." *Cromohs 10*, pp. 1–7.

Meijer Drees, M.E. (1997). *Andere landen, andere mensen. De beeldvorming van Holland versus Spanje en Engeland omstreeks 1650*. Den Haag: Sdu.

Meijer, M.C. (1999). *Race and aesthetics in the anthropology of Petrus Camper (1722–1789)*. Amsterdam: Rodopi.

Meiring, J. (1973). *The truth in masquerade, the adventures of François le Vaillant*. Cape Town: Juta.

"Memorandum voor Gouverneur Janssens over het karakter van met name genoemde boeren." (1932). In E.C. Godée Molsbergen (Ed.), *Reizen in Zuid-Afrika in de Hollandse tijd. Vierde deel. Tochten in het Kafferland 1776–1805*, pp. 210–215. 's-Gravenhage: Nijhoff.

Mentzel, O.F. (1785–1787). *Vollständige und zuverlässige Geographische und Topographische Beschreibung des berühmten und in aller Betrachtung merkwürdigen Afrikanischen Vorgebirges der Guten Hofnung [...]*. 2 vols. Glogau: Christian Friedrich Günther.

Merians, L.E. (2001). *Envisioning the worst: representations of 'Hottentots' in early-modern England*. Newark: University of Delaware Press.

Middelnederlandsch Woordenboek, CD-ROM Middelnederlands (1998). n.p.: Instituut voor Nederlandse Lexicologie/Sdu/Standaard.

Mielke, A. (1993). *Laokoon und die Khoikhoi oder über die Grenzen von Reisebeschreibung und Satire*. Baden-Baden: Koerner.

Miller, C.L. (1985). *Blank Darkness. Africanist discourse in French*. Chicago: University of Chicago Press.

Miller, D.P. (1996). "Joseph Banks, empire and 'centres of calculation' in late Hanoverian London." In D.P. Miller & P.H. Reill (Eds.), *Visions of empire. Voyages, botany and representations of nature*, pp. 21–37. Cambridge: Cambridge University Press.

Mills, S. (1991). *Discourses of difference. An analysis of women's travel writing and colonialism*. London: Routledge.

Milton, J.R. (1989). "Induction before Hume." *British Journal for the Philosophy of Science* 38, pp. 49–74.

Mist, A. de (1835). "Dagverhaal van eene reis naar de Kaap de Goede Hoop, en in de binnenlanden van Afrika door Jonkvr. Augusta Uitenhage deMist, in 1802 en 1803." *Penélopé, of Maandwerk aan het vrouwelijk geslacht toegewijd 8*, pp. 71–127.

—— (1954). *Diary of a journey to the Cape of Good Hope and the interior of Africa in 1802 and 1803*. Cape Town: Balkema.

Mist, J.A. de (1920). "Memorie, houdende de consideratiën en advys van het Departement tot de Indische zaaken, omtrend den voet en wyze, waarop de Regeering van de Caab de Goede Hoop, eventueel zal behooren te worden ingericht." In K.M. Jeffreys & S.F.N. Gie (Ed.), *The memorandum of commissary J.A. de Mist*, pp. 7–158. Cape Town: Van Riebeeck Society.

Mitchell, W.J.T. (1994). *Landscape and power*. Chicago: University of Chicago Press.

Moore, F.C.T. (1969). "Translator's note." In J-M. Degérando (1969), *The observations of savage peoples*, pp. 1–58. London: Routledge & Kegan Paul.

Moravia, S. (1970). *La scienza dell' huomo nel Settecento*. Bari: Laterza.

Mossop, E.E. (1947). *The journals of Brink and Rhenius, being the journal of Carel Frederik Brink of the journey into Great Namaqualand (1761–2) made by Captain Hendrik Hop and the journal of ensign Johannes Tobias Rhenius (1724); transcribed, translated and edited with an introduction, brief lives and footnotes by E.E. Mossop*. Cape Town: Van Riebeeck Society.

—— (Ed.) (1931). *Journals of the expeditions of the Honourable Ensign Olof Bergh (1682 and 1683) and the Ensign Isaq Schrijver (1689). Transcribed and translated into English and edited with a foreword and footnotes by E.E. Mossop*. Cape Town: Van Riebeeck Society.

—— (sa). *Old Cape highways*. Cape Town: Maskew Miller.
Mostert, N. (1992). *Frontiers. The epic of South Africa's creation and the tragedy of the Xhosa people*. London: Jonathan Cape.
Musgrove, B. (1999). "Travel and unsettlement: Freud on vacation." In S. Clark (Ed.), *Travel writing and empire: postcolonial theory in transit*, pp. 31–44. London: Zed.
Muthu, S. (1999). "Enlightenment anti-imperialism." *Social Research* 66, nr. 4, pp. 959–1007.
Muthu, S. (2003). *Enlightenment against empire*. Princeton: Princeton University Press.
"Nachricht von dem Naturalienkabinet des Hrn. le Vaillant (1791)." *Magazin für das Neueste aus der Physik und Naturgeschichte* 7, nr. 2, pp. 93–99.
Naudé, S.D. (1950). "Willem Cornelis Boers." *Argiefjaarboek* 13, nr. II, pp. 355–449.
Newton-King, S. (1999). *Masters and Servants on the Cape Eastern Frontier 1760–1803*. Cambridge: Cambridge University Press.
Nieuw Nederlandsch Biografisch Woordenboek (1911–1937). 10 vols. Leiden: Sijthoff.
Nieuwe algemene beschryving van de Kaap de Goede Hoop (1777). Amsterdam: Conradi.
Nieuwste en beknopte beschryving van de Kaap der Goede-Hope: nevens een dag-verhaal van eenen landtogt, naar het binnenste van Afrika door het land der kleine en groote Namacquas, gedaan op bevel van wylen den Wel Edelen en Gestrengen Heer Ryk Tulbagh, Gouverneur van de Kaap der Goede-Hope, en derzelver ressorte, enz. enz. nevens den Achtbaren Raad van Politie. enz. door een Kommando van LXXXV persoonen (1778). 2 vols. Amsterdam: Schneider.
"Notul of Dagverhaal der Reis en Verrichtingen van President en gecommiteerde Leden uit de Commissie van Veeteelt en Landbouw in de beide Roggevelden, den Hantam, &c." In G.M. Theal (Ed.), *Belangrijke Historische Dokumenten*. Vol. 3, pp. 324–435. Cape Town: Van de Sandt de Villiers.
Ogilvie, J.H. (1962). *A bibliography of Le Vaillant's* Voyages *and* Oiseaux d'Afrique. Johannesburg: Public Library.
Oldfather, C.H. (Ed.). (1935). *Diodorus of Sicily*. Vol. 2. London: Heinemann.
Osterhammel, J. (1998). *Die Entzauberung Asiens: Europa und die asiatischen Reiche im 18. Jahrhundert*. München: Beck.
Outram, D. (1996). "New spaces in natural history." In N. Jardine, J. Secord & E. Spary (Eds.), *Cultures of Natural History*, pp. 249–265. Cambridge: Cambridge University Press.
—— (1999). "On being Perseus: New knowledge, disclosion, and enlightenment exploration." In D.N. Livingstone & C.W.J. Withers (Eds.), *Geography and Enlightenment*, pp. 281–294. Chicago: University of Chicago Press.
Paesie, R. (2002). *Het VOC-fluitschip Stavenisse en de ontdekking van Terra Natal*. Amsterdam: Bataafsche Leeuw.
Pagden, A. (2003). "Human rights, natural rights, and Europe's imperial legacy." *Political Theory* 31, nr. 2, pp. 171–199.
Paravicini di Capelli, W.B.E. (1965). "Reize in de Binnen-Landen van Zuid-Africa gedaan in den Jaare 1803 door W.B.E. Paravicini di Capelli." In W.J. de Kock (Ed.), *Reize in de binnenlanden van Zuid-Africa*, pp. 1–198. Kaapstad: Van Riebeeck Vereniging.
Paterson, W. (1789). *A narrative of four journeys into the country of the Hottentots and Caffraria*. London: Johnson.
Peires, J.B. (1981). *The house of Phalo. A history of the Xhosa people in the days of their independence*. Johannesburg: Ravan.
Penn, N. (1993). "Mapping the Cape: John Barrow and the first British occupation of the Colony, 1795–1803." *Pretexts* 4, nr. 2, pp. 20–43.
—— (2001). "The northern Cape frontier zone in South African frontier historiography." In L. Russell (Ed.), *Colonial frontiers. Indigenous-European encounters in settler societies*, pp. 19–46. Manchester: Manchester University Press.
—— (2005). *The Forgotten Frontier. Colonists & Khoisan on the Cape's Northern Frontier zone in the 18th Century*. Athens/Cape Town: Ohio University Press/Double Storey Books.

Percival, R. (1804). *An account of the Cape of Good Hope.* London: Baldwin.

Philip, J. (1828). *Researches in South Africa, illustrating the civil, moral and religious condition of the native tribes, including journals of the author's travels in the interior, together with detailed accounts of the progress of the Christian missions, exhibiting the influence of Christianity in promoting civilization.* 2 vols. London: Duncan.

Philp, M. (1998). "English Republicanism in the 1790s." *Journal of Political Philosophy* 6, nr. 3, pp. 235–264.

Pickering, A. (1999). "The mangle of practice. Agency and emergence in the sociology of science." In M. Biagioli (Ed.), *The Science Studies Reader*, pp. 372–393. New York: Routledge.

Pire, G. (1956). "Jean-Jacques Rousseau et les relations de voyages." *Revue d'histoire littéraire de la France* 56, pp. 355–78.

Plug, C. (Ed.). (1995). *Robert Jacob Gordon's 'Investigations into meteorology'.* Sl.

Pomian, K. (1990). "Collectors and curiosities. Paris and Venice, 1500–1800." Cambridge: Polity Press.

Porter, D. (1991). *Haunted journeys. Desire and transgression in European travel writing.* Princeton: Princeton University Press.

Pratt, M.L. (1986). "Scratches on the face of the country; or, What Mr. Barrow saw in the land of the Bushmen." In H.L. Gates (Ed.), *'Race', Writing, and Difference*, pp. 138–162. Chicago en London: University of Chicago Press.

Pratt, M.L. (1992). *Imperial eyes. Travel writing and transculturation.* London: Routledge.

Puglisi, C.R. (1996). "Francesco Albani." In *The dictionary of art.* Vol. 1, pp. 533–537. London: Macmillan.

Punt, W. (Ed.) (1976). *Die eerste Europeane in die Nasionale Krugerwildtuin, 1725 (The first Europeans in the Kruger National Park, 1725).* Pretoria: National Parks Board of Trustees.

Quintilian, M.F. (1958). *The Institutio Oratoria of Quintilian.* Vol. I. (H.E. Butler). Cambridge/London: Harvard University Press/Heinemann.

Qureshi, S. (2004). "Displaying Sara Baartman, the 'Hottentot Venus'." *History of Science* 42, pp. 233–257.

Randles, W.G.L. (1959). *L'image de Sud-Est Africain dans la littérature Européenne au XVIième siècle.* Lisboa: Centro de Estudos Históricos Ultramarinos.

Raper, P.E. & Boucher, M. (Eds.) (1988). *Robert Jacob Gordon, Cape Travels, 1777 to 1785.* 2 vols. Houghton: Brenthurst.

"Rapport van het Departement tot de Indische zaaken, nopens het eventueel bestuur van de Kaap de Goede Hoop." (1920). In K.M. Jeffreys & S.F.N. Gie (Ed.), *The memorandum of commissary J.A. de Mist*, pp. 2–6. Cape Town: Van Riebeeck Society.

Raum, J.W. (1997). "Reflections on rereading Peter Kolb with regard to the cultural heritage of the Khoisan." *Kronos* 24, pp. 30–45.

Raven-Hart, R. (1967). *Before Van Riebeeck. Callers at South Africa from 1488 to 1652.* Cape Town: Struik.

Raven-Hart, R. (1971). *Cape Good Hope, 1652–1702. The first fifty years of Dutch colonisation as seen by Callers.* 2 vols. Cape Town: Balkema.

Raynal, G.-T. (1780). *Histoire Philosophique et Politique Des Établissements et du Commerce Des Européens dans les deux Indes. Tome Premier.* A Genève: Chez Jean-Leonard Pellet.

Revel, J. (1991). "Knowledge of the territory." *Science in Context* 4, pp. 133–161.

Rhetorica ad Herennium (1954). (Ed. Harry Caplan). Cambridge: Harvard University Press.

Ringmar, E. (2006). "Audience for a Giraffe: European Expansionism and the Quest for the Exotic." *Journal of World History* 17, no. 4, pp. 375–397.

Roger, J. (1963). *Les sciences de la vie dans la pensée française du XVIIIe siècle.* Paris: Armand Colin.

Roodenburg, H. (sa). *The eloquence of the body. Perspectives on gesture in the Dutch Republic.* Zwolle: Waanders.

Rookmaaker, L.C. (1981). "De Gordon Atlas: achttiende-eeuwse voorstellingen van het Zuidafrikaanse binnenland." *Bulletin van het Rijksmuseum* 29, nr. 3, pp. 123–135.

—— (1983). "The observations of Robert Jacob Gordon (1743–1795) on giraffes (*Giraffa camelopardalis*) found in Namaqualand." *Journal SWA Wissenschaftliche Gesellschaft* 36–37, pp. 71–90.

—— (1989). *The zoological exploration of Southern Africa, 1650–1790.* Rotterdam: Balkema.

—— (1992). "J.N.S. Allamand's additions (1769–1781) to the *Nouvelle Edition* of Buffon's *Histoire Naturelle* in Holland." *Bijdragen tot de Dierkunde* 61, nr. 3, pp. 131–162.

Rookmaaker, L.C. et al. (Eds.) (2004). *François Levaillant and the birds of Africa.* Johannesburg: Brenthurst Press.

Rousseau, J-J. (1755). *Discours sur l'origine et les fondemens de l'inégalité parmi les hommes. Par J.J. Rousseau Citoyen de Genève.* Amsterdam: Marc Michel Rey.

Rousseau, J-J. (1959–1969). *Oeuvres complètes de Jean-Jacques Rousseau.* 4 vols. Paris: Éditions Gallimard.

Rubiés, J-P. (2000). *Travel and ethnology in the Renaissance. South India through European eyes, 1250–1625.* Cambridge: Cambridge University Press.

Ryan, S. (1996). *The Cartographic Eye. How Explorers Saw Australia.* Cambridge: Cambridge University Press.

Saar, J.J. (1662). *Ost-Indianische funfzehenjährigen Kriegs-Dienst und wahrhafte beschreibung.* Nürnberg: Felßecker & Tauber.

Said, E.W. (1991/1978). *Orientalism. Western concepts of the Orient.* Harmondsworth: Penguin.

Sanson, N. (1656). *L'Afriqve, en plvsievrs cartes novvelles, et exactes, &c. en divers traitez de geographie, et d'histoire. Là où sont descrits succinctement, & auec vne belle Methode, & facile. Ses empires, ses monarchies, ses estats, &c. les moevrs, les langves, les religions, le negoce et la richesse de ses peuples, &c. Et ce qu'il y a de plus beau & plus rare dans toutes ses Parties, & dans ses Isles. Par le S. Sanson d'Abbeville, Geographe ordinaire du Roy.* A Paris: chez l'Autheur.

Sarapamadu, S.D. (Ed.) (1956–1957). "An historical relation of Ceylon By Robert Knox." *The Ceylon Historical Journal* 6, nrs. 1–4.

Schapera, I. (1930). *The Khoisan peoples of South Africa: Bushmen and Hottentots*, London: Routledge & Paul.

Schapera, I. & Farrington, B. (1933). *The early Cape Hottentots.* Cape Town: Van Riebeeck Society.

Schepens, G. (1980). *L'Autopsie' dans la méthode des historiens Grecs du Ve siècle avant J.-C.* Verhandelingen van de Koninklijke Academie voor wetenschappen, letteren en schone kunsten van België, *Klasse der Letteren*, 42, nr. 93.

Schöffer, I. & Gaastra, F.S. (1982). "The import of bullion and coin into Asia by the Dutch East India Company in the seventeenth and eighteenth centuries." In M. Aymard (Ed.), *Dutch Capitalism and World Capitalism*, pp. 215–233. Cambridge: Cambridge University Press.

Schouten, W. (1676). *Aanmerckelijcke voyagie gedaan door Wouter Schouten naar Oost-Indien.* Amsterdam: Jacob Meurs & Johannes van Someren.

Schrire, C. (1996). "Native views of western eyes." In P. Skotnes (Ed.), *Miscast. Negotiating the presence of the Bushmen*, pp. 343–353. Cape Town: University of Cape Town Press.

Schutte, G.J. (Ed.) (1982). *Briefwisseling van Hendrik Swellengrebel Jr oor Kaapse sake 1778–1792.* Kaapstad: Van Riebeeck Vereniging.

Seifert, A. (1976). *Cognitio historica. Die Geschichte als Namengeberin der frühneuzeitlichen Empirie.* Berlin: Duncker und Humblot.

Senauze, A. & Horthemels, D. (1686). *Voyage de Siam, des pères jesuites.* Paris: Mort.

Seymour, S. (2000). "Historical geographies of landscape." In B. Graham & C. Nash (Eds.), *Modern historical geographies*, pp. 193–217. Harlow: Pearson Longman.

Shapin, S. (1984). "Pump and circumstance. Robert Boyle's literary technology." *Social Studies in Science 14*, pp. 481–520.

—— (1994). *The social history of truth: Civility and science in seventeenth century England.* Chicago: University of Chicago Press.

—— (1996). *The scientific revolution.* Chicago: University of Chicago Press.

Shapin, S. & Schaffer, S. (1985). *Leviathan and the air-pump. Hobbes, Boyle and the experimental life.* Princeton: Princeton University Press.

Shapiro, B.J. (2000). *A culture of fact. England, 1550–1720.* Ithaca: Cornell University Press.

Shaskan Bumas, E. (2000). "The cannibal butcher shop. Protestant uses of Las Casas's Brevìsima relación in Europe and the American colonies." *Early American Literature 35*, nr. 2, pp. 107–136.

Shepard, O. (1967). *The lore of the unicorn.* London: Allen & Unwin.

Singer, R. (1978). "The biology of the San." In P.V. Tobias (Ed.), *The Bushmen. San hunters and herders of Southern Africa*, pp. 16–32. Cape Town: Human & Rousseau.

Sleigh, D. (1993). *Die buiteposte: VOC-buiteposte onder Kaapse bestuur 1652–1795.* Pretoria: H.A.U.M.

Smith, A.B. (Ed.) (1993). *The Khoikhoi at the Cape of Good Hope: seventeenth-century drawings in the South African Library.* Cape Town: South African Library.

—— (1968). In search of the unicorn. *Africana Notes and News* 18, pp. 91–112.

Smith, A.B. & Pheiffer, R.H. (Eds.) (1992). *Col. Robert Jacob Gordon's notes on the Khoikhoi 1779–80.* Cape Town: *Annals of the South African Cultural History Museum* 5, nr. 1.

Smith, B. (1992). *Imagining the Pacific. In the wake of the Cook Voyages.* New Haven: Yale University Press.

Sparrman, A. (1975–1977). *A voyage to the Cape of Good Hope, towards the Antarctic polar circle round the world and to the country of the Hottentots and Caffers from the year 1772–1776.* 2 vols. Cape Town: Van Riebeeck Society. (Translation of the part on South Africa of: *Resa till Goda Hopps-udden, Södra pol-kretsen, och omkring jordklotet samt till Hottentott- och Caffer-landen åren 1772–76*, Stockholm 1783).

Spary, E.C. (2000), *Utopia's garden. French natural history fromold regime to revolution.* Chicago: University of Chicago Press.

Spicer, J. (1991). "The Renaissance elbow." In J. Bremmer & H. Roodenburg (Eds.), *A cultural history of gesture. From antiquity to the present day*, pp. 84–128. Cambridge: Polity Press.

Spierenburg, P. (1984). *The spectacle of suffering. Executions and the evolution of repression: from a preindustrial metropolis to the European experience.* Cambridge: Cambridge University Press.

Stafford, B. (1984). *Voyage into Substance. Art, Science, Nature, and the Illustrated Travel Account, 1760–1840.* Cambridge (Mass.): MIT Press.

Stagl, J. (1979). "Vom Dialog zum Fragebogen. Miszellen zur Geschichte der Umfrage." *Kölner Zeitschrift für Soziologie und Sozialpsychologie* 31, pp. 611–638.

Stanzel, F.K. (1999). "Zur literarische Imagologie. Eine Einführung." In F.K. Stanzel (Ed.), *Europäische Völkerspiegel. Imagologisch-ethnographische Studien zu den Völkertafeln des frühen 18. Jahrhunderts*, pp. 9–39. Heidelberg: Winter.

Stedman, J.G. (sa). *Narrative of a five years' expedition, against the revolted negroes of Surinam.* London: Johnson & Edwards.

Stepan, N. (1982). *The idea of race in science. Great Britain 1800–1960.* London: Macmillan.

Stocking, G.W. (1968). *Race, culture and evolution. Essays in the history of Anthropology.* New York/London: The Free Press/Collier-Macmillan.

Stone, J.C. (1995). *A short history of the cartography of Africa.* Lewiston: E. Mellen Press.

Streak, M. (1974). *The Afrikaner as viewed by the English 1795–1854.* Cape Town: Struik.

Stresemann, E. (1951). *Die Entwicklung der Ornithologie von Aristoteles bis zur Gegenwart.* Berlin: Peters.

Strother, Z.S. (1999). "Display of the Body Hottentot." In B. Lindfors (Ed.), *Africans on stage. Studies in ethnological show business*, pp. 1–61. Bloomington/Cape Town: Indiana University Press/David Philip.

Suid-Afrikaanse Biografiese Woordeboek (SABW) (1968–1977). 5 vols. Pretoria/Kaapstad: RGN/Tafelberg.

Switzer, L. (1993). *Power and resistance in an African society. The Ciskei Xhosa and the making of South Africa.* Pietermaritzburg: University of Natal Press.

Szalai, M. (1983). *Ethnologie und Geschichte. Zur Grundlegung einer ethnologischen Geschichtsschreibung: mit Beispielen aus der Geschichte der Khoi-San in Südafrika* Berlin: Reimer.

Tachard, G. de (1686). *Voyage de Siam.* Paris: A Paris, Chez A. Seneuze et D. Horthemels.

Taylor, J.G. (1988). *Smeltkroes Batavia: Europeanen en Euraziaten in de Nederlandse vestigingen in Azie.* Groningen: Wolters-Noordhoff.

Theal, G.M. (1898). *Records of the Cape Colony. From December 1796 to December 1799.* Vol. II. N.p.: The Government of the Cape Colony.

—— (1905). *Records of the Cape Colony.* Vol. 31. N.p.: The Government of the Cape Colony.

Thomas, K. (1984). *Man and the natural world. Changing attitudes in England 1500–1800.* Harmondsworth: Penguin.

Thunberg, C.P. (1986). *Travels at the Cape of Good Hope, 1772–1775.* Cape Town: Van Riebeeck Society. (Translation of vol. 4 of *Resan uti kejsardömet Japan, på Java och Ceilon samt hemresan*, Uppsala 1793.)

Tieghem, P. van (1960). *Le sentiment de la Nature dans le Préromantisme Européen.* Paris: Nizet.

Tuck, R. (1999). *The rights of war and peace. Political thought and the international order from Grotius to Kant.* Oxford: Oxford University Press.

Turner, J. (1980). *Reckoning with the beast. Animals, pain, and humanity in the Victorian mind.* Baltimore: The Johns Hopkins University Press.

Tytler, G. (1982). *Physiognomy in the European novel. Faces and fortunes.* Princeton: Princeton University Press.

—— (1995). "Lavater and Physiognomy in English Fiction 1790–1832." *Eighteenth-century Fiction* 7, nr. 3, pp. 293–310.

"Vaillant und Narina." (1791). *Neues Deutsches Museum* 4, pp. 391–394.

Valentyn, F. (1724–1726). *Oud en Nieuw Oost-Indiën.* Dordrecht/Amsterdam: Johannes van Braam/Gerard onder de Linden.

Valentyn, F. (1971–1973). *Description of the Cape of Good Hope with matters concerning it* (Ed. E.H. Raidt). 2 vols. Cape Town: Van Riebeeck Society. (Edition of 'Kaap de Goede Hoop' in F. Valentyn, *Oud en Nieuw Oost-Indiën.*)

Van Berkel, K. (1998). *Citaten uit het boek der natuur. Opstellen over Nederlandse wetenschapsgeschiedenis.* Amsterdam: Bert Bakker.

Van den Boogaart, E. (2000). *Het verheven en verdorven Azië. Woord en beeld in het* Itinerario *en de* Icones *van Jan Huygen van Linschoten.* Amsterdam/Leiden: Spinhuis.

Van der Merwe, J.P. (1926). *Die Kaap onder die Bataafse Republiek 1803–1806.* Amsterdam: Swets & Zeitlinger.

Van Gelder, R. (1993). "Noodzaak of nieuwsgierigheid. Reismotieven van Oostindiëgangers in de zeventiende en achttiende eeuw." *Indische letteren* 8, nr. 2, pp. 50–60.

—— (1997). *Het Oost-Indisch avontuur. Duitsers in dienst van de voc.* Nijmegen: SUN.

Van Grevenbroek, N.N. (1933). "Graevenbroeckii elegans & accurata Gentis Africanae circa Promontorium Capitis Bonae Spei vulgo Hottentotten nuncupata Descriptio Epistolaris Anno MDCLXXXXV." In I. Schapera & B. Farrington (Eds.), *Early Cape Hottentots*, pp. 161–300. Cape Town: Van Riebeeck Society.

Van Hogendorp, G.K. (1801). *Verhandelingen over den Oost-Indischen Handel.* Amsterdam: Weduwe Doll.

—— (1802). *Verhandelingen over den Oost-Indischen Handel, Tweede Stuk. Behelzende eene staatkundige beschouwing van de kolonien benoorden en beöosten, Kaap de Goede Hoop. Met Bylaagen.* Amsterdam: Weduwe Doll.

Van Hogendorp, H. (Ed.) (1876). *Brieven en gedenkschriften van Gijsbert Karel van Hogendorp uitgegeven door zijn jongsten zoon. Derde deel 1788–1813.* 's-Gravenhage: Nijhoff.

Van Reenen, J. (1792) *A journal of a journey from the Cape of Good Hope, undertaken in 1790 and 1791, by Jacob van Reenen, and others of his countrymen, in search of the wreck of the Hounourable East India Company's ship, the Grosvenor; to discover if there remained alive any of the unfortunate sufferers; with additional notes, and a map by Edw. Riou.* London: Nicol. (Translation by Edward Riou.)

Van Maanen, J. (1988). *Tales of the field. On writing ethnography*. Chicago: University of Chicago Press.

Van Stekelenburg, A.V. (2001). "Een intellectueel in de vroege Kaapkolonie: de nalatenschap van Jan Willem van Grevenbroek." *Tydskrif vir Nederlands en Afrikaans 8*, pp. 3–34.

Van Wyk Smith, M. (2003). "The Boers and the Anglo-BoerWar (1899–1902) in the twentieth-century moral imaginary." *Victorian Literature and Culture*, pp. 429–446.

Van Zijl, W.J. (sa). *Van skeepskis na wakis tot boekrak*. Kaapstad: Lux Verbi.

Vermeulen, H. en Kommers, J. (2002). "Introduction. Histories of anthropology in the Netherlands." In H. Vermeulen & J. Kommers (Eds.), *Tales from Academia. History of anthropology in the Netherlands. Part I*, pp. 1–63. Saarbrücken: Verlag für Entwicklungspolitik.

Vermeulen, H.F. (1995). "Origins and institutionalization of ethnography and ethnology in Europe and the USA, 1771–1845." In H.F. Vermeulen & A.A. Roldán (Eds.), *Fieldwork and footnotes. Studies in the history of European anthropology*, pp. 39–59. London: Routledge.

Vermeulen, H.F. (1996). *Taal-, land- en volkenkunde in de achttiende eeuw. Lezing gehouden voor het Oosters Genootschap in Nederland te Leiden op 19 April 1994.* Leiden: Oosters Genootschap in Nederland.

Vogel, J.W. (1690). *Johann Wilhelm Vogels [...] Diarium oder Journal seiner gethanen Reise aus Teutschland nach Holland u. Ost-Indien.* Franckfurth: Augustus Boëtius.

Vogt, P. (1998). "The attitude of eighteenth century German pietism towards Jews and Judaism. A case of Philo-Semitism." *The Covenant Quarterly* 56, pp. 18–32.

Von Bouchenroeder, B.F. (1806). *Reize in de binnenlanden van Zuid-Afrika, gedaan in den jare 1803.* Amsterdam: Mortier Covens.

Voss, A.E. (1979). "Pursuit of the unicorn." *Theoria* 53, *October*, pp. 1–19.

Wheeler, R. (1999). "Limited visions of Africa. Geographies of savagery and civility in early eighteenth-century narratives." In J. Duncan & D. Gregory (Eds.), *Writes of passage: reading travel writing*, pp. 14–48. London: Routledge.

Wheeler, R. (2000). *The Complexion of Race: Categories of Difference in Eighteenth-Century British Culture.* Philadelphia: University of Pennsylvania Press.

Witstein, S. (1969). *Funeraire poëzie in de Nederlandse Renaissance.* Assen: Van Gorcum.

Wolfschmidt, A. (1978). *Magister Peter Kolb. Ein Forscher und Lehrer aus Francken.* Neustadt an der Äsch: private publication.

Woordenboek der Nederlandsche Taal op CD Rom. (2000). n.p.: AND Publishers.

Young, R. (1995). *Colonial Desire: Hybridity in Theory, Culture and Race.* London: Routledge.

Zandvliet, K. (1998). *Mapping for money. Maps, plans and topographic paintings and their role in Dutch overseas expansion during the 16th and 17th centuries.* Amsterdam: Batavian Lion International.

Ziegenbalg, B. (1715). *Bartholomaei Ziegenbalgs Kurze Nachricht von Seiner Reise Aus Ost-Indien nach Europa.* Halle: in Verlegung des Wäysen-hauses.

INDEX

www.ingramcontent.com/pod-product-compliance
Lightning Source LLC
LaVergne TN
LVHW080848300926
849162LV00003B/370